AUSTRALIA'S NUCLEAR POLICY

Australia's Nuclear Policy

Reconciling Strategic, Economic and Normative Interests

MICHAEL CLARKE
Australian National University

STEPHAN FRÜHLING
Australian National University

and

ANDREW O'NEIL
Griffith University, Australia

Routledge
Taylor & Francis Group
LONDON AND NEW YORK

First published 2015 by Ashgate Publishing

2 Park Square, Milton Park, Abingdon, Oxfordshire OX14 4RN
711 Third Avenue, New York, NY 10017

Routledge is an imprint of the Taylor & Francis Group, an informa business

First issued in paperback 2018

British Library Cataloguing in Publication Data
A catalogue record for this book is available from the British Library.

The Library of Congress has cataloged the printed edition as follows:
Clarke, Michael, 1977–
Australia's nuclear policy: reconciling strategic, economic and normative interests / by Michael Clarke, Stephan Frühling and Andrew O'Neil.
pages cm
Includes bibliographical references and index.
ISBN 978-1-4094-4339-1 (hardback)
1. Nuclear weapons—Government policy—Australia. 2. Nuclear nonproliferation—Government policy—Australia. 3. Nuclear energy—Government policy—Australia. I. Frühling, Stephan. II. O'Neil, Andrew. III. Title.
U264.5.A9C53 2015
355.02'170994—dc23

2015004759

ISBN 978-1-4094-4339-1 (hbk)
ISBN 978-1-138-38403-3 (pbk)

Contents

Acknowledgments *vii*
List of Abbreviations *ix*

Introduction: Australia, Nuclear Weapons and Non-Proliferation 1

1 Nuclear Policy: Managing the Dual-Use Challenge of Nuclear Fission 11

2 Exploring Australia's Nuclear Options: Second World War to the Whitlam Government (1945–75) 31

3 Bedding Down the NPT: The Fraser Government (1975–83) 55

4 Nuclear Challenges between Washington and Wellington: The Hawke Government (1983–91) 91

5 New Opportunities and New Dangers: The Keating Government (1991–96) 115

6 Great Power Politics, Proliferation and Terrorism: The Howard Government (1996–2007) 141

7 Consensus and Stagnation in Australian Nuclear Policy: The Rudd-Gillard-Rudd Governments (2007–13) 169

Conclusion: Australian Nuclear Policy in the 20th and 21st Century 187

Bibliography *199*
Index *229*

Acknowledgments

This book represents the culmination of an Australian Research Council (ARC) project examining Australia's nuclear choices. In addition to the financial support of the ARC, we would like to acknowledge the support of Australia's Department of Defence and the Lowy Institute for International Policy, both of which were partners in the project between 2008 and 2012.

We have accumulated a number of important debts in the preparation of this book. We would especially like to thank Hugh White, Richard Brabin-Smith, Allan Behm and Paul Dibb for input at critical stages of the drafting process, and the participants in a seminar at the Strategic and Defence Studies Centre at the ANU for useful suggestions. John Carlson was particularly generous with his time and expertise in providing comments on the final text. We would also like to thank the anonymous referee at Ashgate who provided valuable advice on the review copy.

High value research assistance has been central to the completion of this book during a period where our increasing commitments outside the project meant that time was at a premium. Special thanks goes to Dr Helen Taylor whose research and coordination skills kept us in line and added real value to the final product. Without her support, this book would have taken much longer and may not have been completed. Cathy Moloney, Kalman Robertson and Olivia Cable provided important research assistance at critical stages in the book's evolution. Kathy Bailey and Meegan Thorley deserve thanks for administrative support in managing budgets, booking shuttle services between Canberra and Brisbane, and organizing various meetings. Finally, we would like to thank Brenda Sharp and Stephanie Peake at Ashgate for their professional and highly responsive support, particularly during the final stages of the project.

A few sections in chapters 2–5 draw on Andrew O'Neil's unpublished PhD thesis, 'Constraints and Continuity: The Evolution of Australia's Non-proliferation Policy, 1975–96' (Flinders University, 1999), and Michael Clarke's article 'The Fraser government's uranium decision and the foundations of Australian Nonproliferation Policy: A Reappraisal', *Australian Journal of Politics and History* 58, no. 2 (June 2012), 221–35.

Michael Clarke, Stephan Frühling and Andrew O'Neil
December 2014

List of Abbreviations

AAEC	Australian Atomic Energy Commission
ADF	Australian Defence Force
ALP	Australian Labor Party
ANSTO	Australian Nuclear Science and Technology Organisation
ANZUS	Australia – New Zealand – United States (Treaty)
ASADPO	Australian Strategic Analysis and Defence Policy Objectives
ASNO	Australian Safeguards and Non-Proliferation Office
ASO	Australian Safeguards Office
AUSMIN	Australia-United States Ministerial
BMD	Ballistic Missile Defence
BWR	Boiling Water Reactor
CD	Conference on Disarmament
CTBT	Comprehensive Test Ban Treaty
DFAT	Department of Foreign Affairs and Trade
FTA	Free Trade Agreement
HIFAR	High Flux Australian Reactor
IAEA	International Atomic Energy Agency
ICBM	Intercontinental Ballistic Missile
ICJ	International Court of Justice
ICNND	International Commission on Nuclear Non-proliferation and Disarmament
INFCE	International Nuclear Fuel Cycle Evaluation
NAM	Non-Aligned Movement
NATO	North Atlantic Treaty Organization
NNWS	Non-Nuclear Weapons States
NPG	Nuclear Policy Group
NPR	Nuclear Posture Review
NPT	Non-Proliferation Treaty
NWS	Nuclear Weapons States
OECD	Organization for Economic Cooperation and Development
PHWR	Pressurized Heavy Water Reactor
PNE	Peaceful Nuclear Explosion
PWR	Pressurized Water Reactor
SALT	Strategic Arms Limitation Talks
SEATO	Southeast Asian Treaty Organization
SSAC	State System of Accounting for and Control of Nuclear Material
START	Strategic Arms Reduction Treaty

UMPNER	Uranium Mining, Processing and Nuclear Energy Review
UN	United Nations
UNGA	United Nations General Assembly
USSR	Union of Soviet Socialist Republics

Introduction

Australia, Nuclear Weapons and Non-Proliferation

Australia takes pride in its tradition as an activist middle power with global influence. This tradition is particularly felt in nuclear policy, where Australia has a reputation as a 'champion' of the global nuclear non-proliferation regime. That regime itself is a consequence of the basic 'dual-use' dilemma of nuclear technology, whose promise for economic development inevitably brings with it the dangers of military application. Hence, nuclear policy bridges three related but distinct domains: first, the strategic domain, where nuclear weapons states (NWS) and their allies, including Australia, seek to use the deterrent effect of nuclear weapons to maintain vital security interests; second, the economic domain, which is dominated by the use of nuclear technology to generate power in nuclear reactors; and third, the normative domain, where the ideals of nuclear non-proliferation and nuclear disarmament have been embedded in a global regime, centred on the Treaty on the Non-Proliferation of Nuclear Weapons (NPT). How successive Australian governments have attempted to effectively manage Australian engagement with, and interests in, these three distinct but intimately related domains is the subject of this book.

Existing scholarship on Australia's nuclear policy has generally grappled with this varied history by focusing on discrete aspects of the country's engagement with the three domains.[1] There have been studies on uranium exports and Australia's interest in nuclear technology,[2] on Australian non-proliferation diplomacy,[3] Australian perceptions of the role of nuclear weapons since the

---

1 Rod Lyon, 'Australia: Back to the Future?', in Muthiah Alagappa, ed., *Security in 21st Century Asia* (Stanford, CA: Stanford University Press, 2008). See also Luke Auton, 'Opaque Proliferation: The Historiography of Australia's Cold War Nuclear Weapons Option', *History Compass* 11, no. 8 (2013): 561–72.

2 Alice Cawte, *Atomic Australia 1944–1990* (Sydney: UNSW Press, 1992); Michael Clarke, Stephan Frühling and Andrew O'Neil, eds, *Australia's Uranium Trade: The Domestic and Foreign Policy Challenges of a Contentious Export* (Farnham: Ashgate, 2009); Clarence C. Hardy, *Atomic Rise and Fall: The Australian Atomic Energy Commission 1953–1987* (Peakhurst, NSW: Glen Haven, 1999).

3 Carl Ungerer and Marianne Hanson, eds, *The Politics of Nuclear Non-Proliferation* (St. Leonards, NSW: Allen and Unwin, 2001); Richard Broinowski, *Fact or Fission? The Truth about Australia's Nuclear Ambitions* (Melbourne: Scribe, 2003).

dawn of the atomic age,[4] and Canberra's historical interest in the possibility of acquiring nuclear weapons.[5] In their focus on specific aspects of nuclear policy, most of these studies reflect not only the specific research questions (and in some cases, policy advocacy) of their authors, but also governmental divides between foreign affairs and diplomats, the defence establishment, the nuclear research institutes, and mining and industry.

Moreover, many existing studies of nuclear policy have tended to treat their subject in relative isolation from the wider development of international nuclear affairs – a discretion and luxury that is seldom granted in practice to governments, strategists and diplomats. Since the 1970s, the manner in which Australian governments have attempted to mediate the country's often divergent interests across the strategic, economic and normative domains of nuclear policy has been crucially shaped by the policies of other countries. For example, during the Fraser government (1975–83), international developments within the strategic and economic domains – including India's 'peaceful nuclear explosion' of 1974 and the shift towards 'breeder' reactor technology – interacted to inform that government's nuclear safeguards and uranium export policy. Such a dynamic has in fact been a recurring theme in Australian nuclear policy since that time and it is possible to discern a number of critical junctures in which dynamics at the global level have converged to produce shifts in Australia's approach.

In contrast, this book argues for an understanding of nuclear policy that extends across all of these fields. Unlike most of the existing secondary literature, the focus here is on examining how and why successive Australian governments have engaged with nuclear issues, particularly from a security perspective. Framing the topic in this manner means that what might initially appear as inconsistent positions that hinder progress towards specific policy objectives – such as, for example, the maintenance of US extended nuclear deterrence while simultaneously seeking nuclear disarmament – can in fact be seen as rational approaches to the broader goal of managing strategic risks in general, and not only those arising from the existence of nuclear weapons. Behind this seemingly academic question are deep-held assumptions about what are and should be the ends and means of nuclear policy, and even the

---

4 Christine Leah and Rod Lyon, 'Three Visions of the Bomb: Australian Thinking about Nuclear Weapons and Strategy', *Australian Journal of International Affairs* 64, no. 4 (2010): 449–74; Christine Leah, 'US Extended Deterrence and Nuclear Order: An Australian Perspective', *Asian Security* 8, no. 2 (2012): 93–114.

5 Jim Walsh, 'Surprise Down Under: The Secret History of Australia's Nuclear Ambitions', *The Nonproliferation Review* 5, no. 1 (1997): 1–20; Wayne Reynolds, *Australia's Bid for the Atomic Bomb* (Melbourne: Melbourne University Press, 2000); Jacques Hymans, 'Isotopes and Identity: Australia and the Nuclear Weapons Option, 1949–1999', *The Nonproliferation Review* 7, no. 1 (2000): 1–23.

purpose of nuclear weapons, about which there is rarely complete agreement among Australian policy makers, advocates or analysts. How to reconcile these differences is an important part of the history, practice and future of Australia's nuclear policy.

This book will analyze Australia's nuclear policy from a historical perspective – focusing on the country's strategic engagement with nuclear deterrence, its economic role as a uranium exporter and its participation in the normative non-proliferation system – with a particular emphasis on the changing international context within which policy has developed. It will examine what were perceived as main policy challenges over time and what diplomatic, political and technical contributions Australia sought to make in addressing these challenges. While the book analyzes Australia's increasing engagement in the nuclear fuel cycle after 1945, it devotes particular attention to evaluating the policy approaches between 1975 and 2013, encompassing the Fraser, Hawke, Keating, Howard and Rudd-Gillard governments. The book concludes with a systematic appraisal of the key findings flowing from the analysis and what this means for the future of Australia's nuclear policy in the 21st century.

Australian Foreign Policy Traditions and Australian Nuclear Policy

Since the early 1970s it has been a central and bipartisan goal of Australian foreign policy to limit the further proliferation of nuclear weapons in the international system. The proposition that it is both right and necessary for Australia to contribute to global efforts to halt nuclear proliferation is generally perceived as self-evident, and there are few other policy issues that enjoy such bipartisan support as nuclear non-proliferation. Successive Australian governments did not think it would be best to achieve this aim by disengaging from the development and use of nuclear technology itself, or the nuclear aspects of the US alliance – in the way that New Zealand would do during the 1980s. This bipartisan consensus stems from the striking of a nuclear 'grand bargain' in the late 1970s, whereby Australia would export uranium oxide (or 'yellowcake') to non-nuclear weapons states only if they were signatory to the NPT. This policy was based on the judgment that only by becoming a substantial player in the international nuclear fuel cycle could Australia become a significant contributor to global nuclear non-proliferation efforts as well. It is with the development of this 'grand bargain' under the Fraser government that the detailed analysis of the book begins.

Over the past two decades, however, there has been a vigorous debate in the academic literature that raises doubts about the notion that Australian foreign policy has exhibited genuine bipartisanship on the big questions of how the country should relate to the world. In lieu of foreign policy bipartisanship, this literature identifies the contestation and interaction of various strands or 'traditions' of foreign policy as a key explanatory variable for the trajectory of Australian foreign policy since the Second World War.[6] Some of this literature has a revisionist quality and seeks to challenge what its protagonists identify as a prevailing 'orthodox' interpretation of Australian foreign policy.[7] This orthodox interpretation, it has been argued, was 'constructed' during the sustained domestic political dominance of the Australian Labor Party (ALP) under prime ministers Bob Hawke (1983–91) and Paul Keating (1991–96), and conceives of Australian foreign policy proceeding through a number of stages since 1945 from 'dependency' for security on 'great and powerful friends' to a more self-assured 'middle power diplomacy'.

In this schema, Australian foreign policy from 1945 to at least the mid-1970s is defined by the country's struggle to assert its own foreign policy identity. This struggle for self-realization was constrained by the combined weight of the realities of Australia's geography, relative power and culture and the dynamics of the Cold War in Asia that compelled it to seek the protection of its 'great and powerful' friends, the United Kingdom and the United States.[8] This dependence for security on the United Kingdom and the United States was then crucially 'broken' in the early 1970s by the combined weight of domestic political change and the perceived decline in the capability of Australia's great and powerful friends to protect it in the wake of the United Kingdom's retreat from 'east of Suez' and the US withdrawal from Vietnam.[9] Thereafter Australian foreign

---

6 Trevor Matthews and John Ravenhill, 'Bipartisanship in the Australian Foreign Policy Elite', *Australian Outlook* 42, no. 1 (1988): 9–20; Michael Wesley and Tony Warren, 'Wild Colonial Ploys? Currents of Thought in Australian Foreign Policy', *Australian Journal of Political Science* 35, no. 1 (2000): 9–26; David M. Jones and Mike L. Smith, 'Misreading Menzies and Whitlam: Reassessing the Ideological Construction of Australian Foreign Policy', *The Round Table: The Commonwealth Journal of International Affairs* 89, no. 355 (2000): 387–406; Owen Harries, 'Punching above Our Weight?', Boyer Lecture, 21 December 2003, available at http://www.abc.net.au/radionational/programs/boyerlectures/lecture-6-punching-above-our-weight/3459850 [accessed 10 June 2013]; David M. Jones and Andrea Benvenuti, 'Tradition, Myth and the Dilemma of Australian Foreign Policy', *Australian Journal of International Affairs* 60, no. 1 (2006): 103–24.

7 See for example Jones and Benvenuti, 'Tradition, Myth and the Dilemma of Australian Foreign Policy'.

8 David McLean, 'Australia in the Cold War: A Historiographical Review', *The International History Review* 23, no. 2 (2001): 299–321.

9 Meg Gurry, 'Identifying Australia's "Region": From Evatt to Evans', *Australian Journal of International Affairs* 49, no. 1 (1995): 17–31; Allan Gyngell and Michael Wesley, *Making Australian Foreign Policy*, 2nd edition (Cambridge: Cambridge University Press, 2007); David Reynolds, 'Empire, Region, World: the International Context of Australian Foreign Policy since 1939', *Australian Journal of Politics and History* 51, no. 3 (2005): 346–58.

policy is understood to have been increasingly defined by a commitment to regional engagement with its Asian neighbours through energetic 'middle power diplomacy' and a predilection for multilateralism.[10]

This broader, orthodox narrative has also had a major influence on how the evolution of Australian nuclear policy has been interpreted – and Australia's shift from reluctant to enthusiastic support for non-proliferation is in turn part of the narrative itself. This is because Australia's approach to nuclear weapons and non-proliferation is also split between two relatively clear phases that correlate with those of the 'middle power' narrative. The first phase, from the closing years of the Second World War to the early 1970s, was defined by four major themes: efforts to harness the strategic and economic potential of atomic energy to Australian needs; an ambivalence towards emerging arms control initiatives; generally unquestioned support for the nuclear programs of its allies; and the unfulfilled economic potential of Australia's uranium resources.[11] In this interpretation Australia's approach to nuclear issues is taken to be derivative of its prevailing security anxieties and political culture, which were defined by an unsophisticated view of the Cold War in Asia and a 'derivative' and 'conservative' Anglo-Celtic heritage.[12] Jacques Hymans explained Australian governments' interest in nuclear weapons with the 'oppositional identities' of Australian leaders at the time, although it was only 'oppositional nationalists' like John Gorton who remained sympathetic to a sovereign Australian nuclear capability, rather than relying on UK and US deterrence.[13]

In contrast, the second phase, beginning with Australia's ratification of the NPT as a non-nuclear weapons state in 1973 under the Labor government of Prime Minister Gough Whitlam has been defined by a strong and activist commitment to the international non-proliferation regime. Australia's reorientation of its approach to nuclear issues after 1972, towards multilateralism and commitment to international agreements, is taken to be a 'natural' expression of the Whitlam government's reversal of the country's perceived relationship to its immediate region. Thus Meg Gurry argues that Australia's turn towards 'multilateralism' and 'regionalism' was the product of a re-orientation of the country's identity from one seeking security *from* its region to one that conceived of itself as 'an *independent*, medium-sized power located *in* the Asia-Pacific region'.[14] Hymans likewise suggests that Australian

10 Carl Ungerer, 'The "Middle Power" Concept in Australian Foreign Policy', *Australian Journal of Politics and History* 53, no. 4 (2007): 538–51.

11 Reynolds, *Australia's Bid*; Cawte, *Atomic Australia*; Walsh, 'Surprise Down Under: The Secret History of Australia's Nuclear Ambitions'.

12 Gurry, 'Identifying Australia's "Region": From Evatt to Evans', 21–5.

13 Hymans, 'Isotopes and Identity: Australia and the Nuclear Weapons Option, 1949–1999'.

14 Gurry, 'Identifying Australia's "Region": From Evatt to Evans', 17.

governments from Whitlam on became 'non-oppositional', and hence saw nuclear weapons as less relevant for their security.[15]

In this second phase, Australia became a 'champion' of non-proliferation through diplomatic activism in international and regional non-proliferation forums, and Australian policy thereafter rested on three pillars: first, using exports of its uranium reserves to establish its status as a non-nuclear weapons state with a significant role in the international nuclear fuel cycle and using its export policy to strengthen non-proliferation through strict export controls; second, using its commitment to export controls to extend the coverage of the non-proliferation regime and as part of its activism to expand the membership of the NPT, and add supplementary agreements such as the Comprehensive Test Ban Treaty (CTBT) to that regime; and third, supporting US extended nuclear deterrence while renouncing the domestic nuclear weapons option.

The influence of this logic reached its apogee with Australia's sponsorship of the Canberra Commission on the Elimination of Nuclear Weapons in 1995. In its aspirations and method, Australian non-proliferation diplomacy arguably displayed those traits of a 'middle power' that

> manifests in altruistic foreign policy behaviour broadly defined as activism on behalf of global causes that reflect the interests of the international community as a whole rather than an individual state's own national or alliance interests.[16]

According to Hanson and Ungerer, Australia's sponsorship of the Canberra Commission was possible because it was not only 'a committed member of the Western alliance' but also 'far enough away, both strategically and psychologically, from the centres of power to distance itself from the nuclear posturings of the Cold War era'.[17] Moreover, 'by never seriously pursuing a nuclear weapons capability for itself' Australia could claim a position of leadership as a 'committed non-proliferator urging non-proliferation and disarmament at the global level'.[18] Australia's systemic role as a 'status quo-oriented, satisfied, system-affected middle power' informed its preference for establishing and maintaining a 'rules based' international order. Thus Australia since the late 1970s has become a prominent member of a group of states that Hanson has termed the 'advocacy states' (Australia, Japan, Canada, Ireland, Sweden, Norway and New Zealand)

---

15 Hymans, 'Isotopes and Identity: Australia and the Nuclear Weapons Option, 1949–1999'.

16 David A. Cooper, 'Challenging Contemporary Notions of Middle Power Influence: Implications of the Proliferation Security Initiative for "Middle Power Theory"', *Foreign Policy Analysis* 7, no. 3 (2011): 317–36, 321.

17 Marianne Hanson and Carl Ungerer, 'The Canberra Commission: Paths Followed, Paths Ahead', *Australian Journal of International Affairs* 53, no. 1 (1999): 5–17, 11.

18 Ibid., 11.

within the non-proliferation regime, who share the 'strong conviction that while nuclear weapons cannot be uninvented, their possession and use – and eventual elimination – can nevertheless be managed by strong norms and regulatory institutions'.[19]

Security, Alliance and Australian Engagement with Nuclear Weapons

The 'middle power' explanation for the long endurance of the 'grand bargain' raises an important question. What accounts for the continuity of the 'middle power' rhetoric and themes in the statements on nuclear issues of successive Prime Ministers and Foreign Ministers of both Labor and Coalition governments? Has this been the result of the 'internalization', in Finnemore and Sikkink's terms, of the norms of non-proliferation and nuclear disarmament by Australian policy-makers since the mid-1970s?[20] At first glance this argument would seem to have purchase given that governments of both major political persuasions have upheld the basic foundations and objective of the 'grand bargain' since the late 1970s, and made notable contributions to international non-proliferation diplomacy.

Although not disputing that Australia's self-conception as a 'responsible middle power' has influenced its approach to nuclear policy in significant ways, this book argues that it has been only one factor influencing policy. Michael Wesley, for instance, highlights the important role played by international security interests in determining Australia's normative commitments. He argues that '[a] sense of isolation … twinned with a sense of diminutive size' has led to the conviction amongst Australians that 'they are unable to provide absolute security for themselves'.[21] This accounts for what Wesley terms 'the greatest continuity in Australian foreign policy history' – the reliance on alliance with great powers for security. This reliance on great power allies, however, produced not only 'dependence' for security on its great power protector but also the linkage of Australian security in the minds of policy-makers with that of the international system itself:

> Aware that a challenge to the global position of its great power ally could have ramifications for regional stability and Australia's specific security interests,

---

19 Marianne Hanson, 'The Advocacy States: Their Normative Role before and after the U.S. Call for Nuclear Zero', *The Nonproliferation Review* 17, no. 1 (2010): 71–93, 73.

20 Martha Finnemore and Kathryn Sikkink, 'International Norm Dynamics and Political Change', *International Organization* 52, no. 4 (1998): 887–917.

21 Michael Wesley, 'Setting and Securing Australia's National Interests: The National Interest as Security', in Ian Marsh, ed., *Australia's Choices: Options for a Prosperous and Fair Society* (Sydney: UNSW Press, 2003), 180.

> Australian governments have consistently voiced strong support for their great power ally's efforts in support of global order.[22]

Consistent with this approach, the most influential Australian view on nuclear weapons throughout the last six decades was what Rod Lyon and Christine Leah dubbed a 'Menzian' one. It values the international order supported by nuclear weapons in the hands of the great powers, and regards proliferation to other countries, including Australia, as threatening the global stability that the country's security ultimately depends on. Contrasting views, which focus more on the immediate security benefits of a sovereign nuclear capability, or reject nuclear weapons completely in favour of disarmament, were also enduring features of the Australian debate throughout the whole nuclear age, but could never successfully challenge the 'Menzian' majority.[23]

Inherent in the 'Menzian' view was a focus on the role of nuclear weapons in international order. Throughout the Cold War, Australia was conscious of the importance of nuclear weapons for the defence of the West. However, it was also removed from the centres of Cold War confrontation, so that these concerns were not immediately relevant to its own defence. But compared to many US allies in the Northern Hemisphere, it was not only Australia's involvement with nuclear strategy that was relatively removed and indirect. Australia also only has very limited civilian nuclear industry – unlike Canada, for example, it never developed nuclear reactors to capitalize on its uranium ore deposits. Even its uranium exports never fully lived up to their potential, as Australia's market share continues to lag well behind its share of worldwide deposits. Hence, for Australia it was the promise – rather than the reality – of strategic or economic benefit from involvement in the nuclear fuel cycle that influenced nuclear policy over several decades.

Although it ultimately eschewed nuclear power production, Australia did build up a sizeable nuclear research establishment in the 1950s and 1960s. The Australian Atomic Energy Commission operated uranium centrifuges in the 1970s and into the 1980s, and ANSTO to this day researches technologies that are of relevance for reactor and fuel cycle design. In addition to the popular narrative of Australia as a champion of non-proliferation diplomacy, Australia also has a distinct if less prominent tradition of technical contributions to practical problems that arise in non-proliferation and the safeguarding of nuclear facilities, including Australia's Safeguards Support Program cooperation with the IAEA, participation in the Hexapartite Safeguards Project, or the analysis of IAEA samples by ANSTO laboratories. And whereas many of these practical activities have global application, Australia has also helped countries

22 Ibid.

23 Leah and Lyon, 'Three Visions of the Bomb: Australian Thinking about Nuclear Weapons and Strategy'.

in the immediate region implement nuclear safeguards, export controls, and regulatory and safety measures.

One of the central aims of this book is thus to demonstrate that while the tenor and activism of Australian non-proliferation diplomacy may speak to the 'middle power' discourse of 'good international citizenship', and while this narrative of Australia as a champion of non-proliferation is not wrong, Australia's contributions were more nuanced and broader than a focus on the efforts of Australian diplomats alone would suggest. Because Australia's main benefit from the civilian use of nuclear technology relied on the spread of nuclear power to other countries, the dual-use dilemma was, and is, particularly acute and important in its nuclear policy debates. Understanding nuclear technology was an important element in being able to contribute to non-proliferation and safeguards in practice, but rests on a very narrow basis in terms of funding, personnel and expertise. The international conditions in which Australia was able to make contributions to the development of the non-proliferation regime were a major factor in its motivation and ability to do so, but they have undergone significant change. Australian nuclear policy of the last 70 years has evolved significantly over time, but it has continued to draw on similar ideas and resources. The same will also be true for the future of Australia's nuclear policy in the 21st century.

***

This book examines how and why Australia has made key choices in relation to the economic, strategic and normative dimensions of the nuclear fuel cycle since the 1970s, when the Fraser government laid the foundation for Australia's contemporary nuclear policy. The book comprises seven chapters and is chronological in order. In Chapter 1, we proceed to define what we mean by 'nuclear policy'. The key argument put forward in this chapter is that nuclear policy is made up of many dimensions that bear directly on major strategic, economic and normative questions. The second chapter discusses the postwar development of Australia's engagement in the nuclear fuel cycle from 1945 to 1975. It focuses on the debate over whether Australia should acquire a nuclear force, the relationship between national development and nuclear power, and the relationship between nuclear exports and international non-proliferation objectives. Chapter 3 analyzes the Fraser government's (1975–83) approach to nuclear policy, and here we argue that the so-called 'grand bargain' enunciated in 1977 has essentially set the parameters of Australian nuclear policy since. In Chapter 4, the Hawke government's (1983–91) nuclear policy is discussed. A major theme of this chapter is that the salience of nuclear weapons for Australia's security gradually declined as the end of the Cold War approached. Chapter 5 analyzes the Keating government's (1991–96) nuclear policy, which we argue reflected a more permissive environment for Australian efforts to promote nuclear arms control in international forums. On the other hand, Australian concerns about the future of the US presence in Asia after the Cold War manifested itself in a desire for reassurance with respect to extended nuclear deterrence. In Chapter 6, we assess the Howard government (1996–2007) and argue that its time in office was characterized by a willingness to review some of the key tenets underpinning the 'grand bargain' logic, including approving uranium sales to non-NPT member India. Finally, Chapter 7 examines the Rudd-Gillard-Rudd governments (2007–13). While the first Rudd government (2007–10) reversed key nuclear policy decisions of the Howard government, by the time Labor lost the 2013 election, Australian nuclear policy had returned to a high degree of bipartisanship. In the concluding chapter, we discuss the major themes to emerge from the preceding analysis and draw some lessons for the future of Australian nuclear policy.

Chapter 1
Nuclear Policy: Managing the Dual-Use Challenge of Nuclear Fission

The term 'nuclear policy' is not as straightforward as it may appear. In public debate and government organization, it is immediately obvious what is meant by 'health policy', 'education policy' or 'defence policy'. All of these terms are well established in popular and official use; in the Australian context all of them are aligned with government departments that exist to develop and implement them; and all of them are, grammatically, compound nouns. Not so 'nuclear policy'. Its use is, as we shall see, inconsistent. It is rarely obvious which part of government is responsible for the development and implementation of nuclear policy as a whole; and as a term, it combines adjective and noun. 'Nuclear policy' shares the last of these traits with the more common term of 'foreign policy' – but whereas policy can be meaningfully said to be 'foreign' or 'domestic', it does not make much sense for it to be 'nuclear' itself. Nor does 'nuclear policy' refer to the nucleus of policy. 'Nuclear policy' is a term that is missing a noun.

What is 'Nuclear Policy'?

What might this noun be? There are several possibilities. The US Department of Defense, for example, has a 'Deputy Assistant Secretary of Defense for Nuclear and Missile Defense Policy', whose responsibilities include nuclear weapons and nuclear arms control. In the US defence context, 'nuclear policy' is thus generally used as shorthand for 'nuclear weapons policy'. This is the case with the use of the term by the Department of State,[1] in the academic defence debate,[2] and in the programs of the Center for Strategic and International Studies, a major US think tank.[3] Other US organizations, including the Heritage Foundation,

1 See for example the speech by Frank A. Rose, Deputy Assistant Secretary, Bureau of Arms Control, Verification and Compliance: Frank A. Rose, 'The Future of U.S. Nuclear Policy' (The Pryzbyla Center, Catholic University of America, Washington, DC, 9 April 2013), available at http://www.state.gov/t/avc/rls/2013/207360.htm [accessed 20 June 2013].

2 See for example Keith B. Payne and Colin S. Gray, 'Nuclear Policy and the Defensive Transition', *Foreign Affairs* 62, no. 4 (1984): 820–42.

3 See 'Nuclear Issues Project', the Center for Strategic and International Studies, available at http://csis.org/program/nuclear-policy-news [accessed 20 June 2013].

Council on Foreign Relations, Natural Resources Defense Council or Union of Concerned Scientists, use the longer 'nuclear weapons policy', or both.[4]

Reading 'nuclear policy' as 'nuclear weapons policy' or 'nuclear non-proliferation policy' would, however, confound those who consult the United Kingdom's Nuclear Industry Association[5] or the 'World Nuclear News'.[6] Both employ the term 'nuclear policy' as shorthand for 'nuclear power policy' or 'nuclear energy policy'. Again, this is a use of the term 'nuclear policy' that also extends to the relevant academic literature.[7]

Then, finally, there are those cases where 'nuclear policy' is used to refer to all of the above: nuclear weapons, arms control, non-proliferation, energy, as well as – in more recent years – nuclear safety and security. 'Nuclear policy' programs of the Lowy Institute for International Policy in Australia,[8] and the US Carnegie Endowment for International Peace fall into this category.[9] And this is also the conception that underpins this book. At first glance, this seems an unwieldy use of the term – depending on the particular issue involved, 'nuclear policy' could then be integral to broader defence, foreign, energy, industry, science, resources, mining and even health policy, with a commensurately broad range of governmental departments and agencies involved. Yet, our argument is that this breadth and the connection between different areas are real, and indeed

---

4 William J. Perry and Brent Scowcroft (chairs), 'U.S. Nuclear Weapons Policy', *Independent Task Force Report* no. 62 (Washington, DC: Council on Foreign Relations Press, 2009); Michaela Dodge and Baker Spring, 'Bait and Switch on Nuclear Modernization Must Stop', *Backgrounder* no. 2755 (Washington, DC: Heritage Foundation, 2013); 'Nuclear Weapons Policy', Natural Resources Defense Council, available at http://www.nrdc.org/nuclear/nuguide/nwipoli.asp [accessed 20 June 2013]. George Washington University's 'Nuclear Policy Talks' are a relative outlier, as they focus on nuclear non-proliferation and disarmament policy. See 'Nuclear Policy Talks', George Washington University Elliott School of International Affairs, available at http://elliott.gwu.edu/news/events/nuclear-policy-talks.cfm [accessed 20 June 2013].

5 See 'Nuclear Policy in the UK – Where are we now?', Nuclear Industry Association, available at http://www.niauk.org/nuclear-policy-in-the-uk [accessed 20 June 2013].

6 'Nuclear Policies', World Nuclear News, available at http://www.world-nuclear-news.org/sectionhub.aspx?fid=800 [accessed 20 June 2013].

7 See for example Måns Nilsson, 'The Role of Assessments and Institutions for Policy Learning: A Study on Swedish Climate and Nuclear Policy Formation', *Policy Sciences* 38, no. 4 (2005): 225–49.

8 'Nuclear Policy Centre', Lowy Institute for International Policy, available at http://m.lowyinstitute.org/programs-and-projects/projects/nuclear-policy-centre [accessed 20 June 2013].

9 'About the Carnegie Nuclear Policy Program', Carnegie Endowment for International Peace, available at http://carnegieendowment.org/programs/npp/?fa=about [accessed 20 June 2013].

that they are the main challenge for the development and implementation of nuclear policy in practice. Indeed, despite the stove-piped nature of government administration, Australian governments have in the past explicitly considered nuclear policy as an area that by its nature spans both many departments (e.g. when the Hawke Cabinet considered the scope for savings in administering 'nuclear policy'),[10] and constitutional powers (e.g. when the Fraser Cabinet considered the basis for a review of nuclear policy legislation).[11]

What, then, is the missing noun in 'nuclear policy'? In this expansive perspective, 'nuclear fission policy' may perhaps best capture the underlying theme. The common basis of nuclear weapons, arms control and non-proliferation, and of nuclear power production, safety and security is that they ultimately all relate to the use and control of the fission of atom cores of uranium-235 and plutonium-239.[12] In 1976, the then US President Gerald Ford began his 'Statement on Nuclear Policy' with a discussion of the imperative to reconcile the inherent promise and the inherent danger of nuclear fission, in a passage that is as valid today as it was then:

> We have known since the age of nuclear energy began more than 30 years ago that this source of energy had the potential for tremendous benefits for mankind and the potential for unparalleled destruction. On the one hand, there is no doubt that nuclear energy represents one of the best hopes for satisfying the rising world demand for energy with minimum environmental impact and with the potential for reducing dependence on uncertain and diminishing world supplies of oil. On the other hand, nuclear fuel, as it produces power also produces plutonium, which can be chemically separated from the spent fuel. The plutonium can be recycled and used to generate additional nuclear power, thereby partially offsetting the need for additional energy resources. Unfortunately – and this is the root of the problem – the same plutonium produced in nuclear power plants can, when chemically separated, also be used to make nuclear explosives. The world community cannot afford to let potential nuclear weapons material or the technology to produce it proliferate uncontrolled over the globe. The world community must ensure that production and utilization of such material by any nation is carried out under the most stringent security conditions

10 Cabinet Decision No. 8344 (AER), 31 July 1986, NAA A14039, 4155.

11 Cabinet Decision No. 3654, 16 August 1977, NAA A12909, 1568.

12 Uranium-235 and plutonium-239 are fissile materials, because they can sustain a chain reaction. There are other fissile materials in the actinide group, but with the possible exception of uranium-233 they are not of practical importance. Nuclear reactions in both bombs and reactors also include the fission of other, non-fissile but 'fissionable' materials, such as uranium-238. Many atoms fission spontaneously and thereby release radioactivity which can have useful applications in industry, but 'fission' herein is used to refer to fission chain reactions in bombs or reactors, unless the contrary intention is apparent.

> and arrangements. Developing the enormous benefits of nuclear energy while simultaneously developing the means to prevent proliferation is one of the major challenges facing all nations of the world today.[13]

In principle, the dual-use challenge of nuclear technology is not unusual. Almost any technology from knives or guns to chemical and biological agents can be used for military as well as civilian applications. But whereas nuclear policy has developed to deal with the dual-use dilemma of nuclear fission, there is no equivalent 'oxidization policy' that would do the same for conventional explosives. Nuclear policy is the only policy area that concerns humankind's interaction with a physical or chemical phenomenon.

In this respect as well, nuclear policy is thus quite distinct from other areas of public policy. The importance of the state and government is far greater in the development and management of almost any application of nuclear fission than it is in other technologies, with the possible exception of spacefaring. This is not a mere consequence of the particular destructive power of nuclear weapons, which might have motivated governments to maintain close control of all matters nuclear. Rather, the use of nuclear fission – similar to early space technology – was and remains of a scale and difficulty that the active support of governments, in one form or another, remains a necessary condition. Because of the destructive power and the economic potential of nuclear fission, states have the motivation to make nuclear policy; because of the nature of the technology and its application, they have the opportunity to do so; but because of the wide range of considerations, objectives and actors involved – from defence to economic, science, health and diplomacy – the whole that is nuclear policy is also often overshadowed by its parts.

Overall, nuclear policy consists of three distinct domains: that of strategy; that of economics; and that of nuclear norms. All are inseparably linked by the inherent dual-use challenge of nuclear technology and fissile materials. This chapter will now introduce each domain separately, before demonstrating how the role of the state, its security and technological expertise are the common foundations of nuclear policy in all three of them.

The Strategic Domain of Nuclear Policy

It is to the military application of atomic energy that humankind owes its entry into the nuclear age. In early 1939, Lise Meitner and Otto Frisch identified atomic fission in experimental results of Otto Hahn and Fritz Strassmann. In August of the same year, Albert Einstein wrote his now-famous letter to US

13 Gerald R. Ford, 'Statement on Nuclear Policy', 28 October 1976, online at Gerhard Peters and John T. Woolley, *The American Presidency Project*, available at http://www.presidency.ucsb.edu/ws/?pid=6561 [accessed 20 June 2013].

President Franklin D. Roosevelt, in which he stated that 'it is conceivable … that extremely powerful bombs of a new type may thus be constructed'.[14] Throughout the Second World War, the United Kingdom and United States saw themselves in a race against Germany to harness the power of the atom. By May 1945, it was clear that the German effort had been far less advanced than feared,[15] and the Manhattan project led to the first ever atomic explosion at Alamogordo on 16 July 1945.

The Second World War provided not only the motivation to create atomic bombs, but also the initial doctrinal and technical basis to make military use of their unprecedented power: strategic bombing by means of US long-range bombers. As such, the attacks on Hiroshima and Nagasaki on 6 and 9 August 1945 were different in kind, but not in their strategic logic, intent and scale from the conventional attacks on German and Japanese cities that had preceded them. And yet, they remain the only use of nuclear weapons in anger, and inseparably linked to the image of indiscriminate urban destruction. To make use of nuclear weapons, countries thus do not only have to overcome the technological hurdles of building them – a task that arguably has become easier as the level of general manufacturing technology has improved. They also need to find a credible strategy that can link the threat of large-scale nuclear destruction to the achievement of their national objectives. Although the former remains a significant task, the latter arguably proves just as, if not even more difficult.

Early imagination about the strategic effect of atomic bombs focused on the threat from nuclear aggression. Thankfully it is rare that leaders with genocidal intentions rise to command the full resources of a modern industrial country. But whereas most nuclear strategy in later decades focused on deterrence, it was the spectre of a 'nuclear Pearl Harbour' that preoccupied early thinking about the dangers of 'the bomb'.[16] But even if nuclear weapons are not used directly, the ability to deter large-scale retaliation means that dissatisfied powers armed with nuclear weapons can engage in more provocative behaviour than they otherwise would. What is often referred to as the 'stability-instability paradox' has been used to explain Pakistan's initiation of the 1999 Kargil War, after it had demonstrated its nuclear capabilities in tests the previous year.[17]

14 See Glenn Elert, 'Albert Einstein's Letters to President Franklin Delano Roosevelt', *E-World*, available at http://hypertextbook.com/eworld/einstein.shtml#first [accessed 2 August 2010].

15 Jeffrey T. Richelson, *Spying on the Bomb: American Nuclear Intelligence from Nazi Germany to Iran and North Korea* (New York: W.W. Norton & Company, 2006), 50–61.

16 Bernard Brodie, 'Implications for Military Policy', in Bernard Brodie, ed., *The Absolute Weapon* (New York: Harcourt, Brace and Company, 1946), 76.

17 See for example Michael Krepon, 'The Stability-Instability Paradox, Misperception, and Escalation-Control in South Asia', in Raaq Dossani and Henry S. Rowen, eds, *Prospects for Peace in South Asia* (Stanford, CA: Stanford University Press, 2005).

Over the 30 years following the Second World War, the state of the art in the technology of nuclear weapons, and the means to deliver them, advanced in leaps and bounds. The development of miniaturized fission warheads in the early 1950s opened up a whole range of tactical applications. Thermonuclear warheads, which combine nuclear fission of plutonium or uranium with fusion of light atomic cores of deuterium and lithium in a staged warhead design, became available in the mid-1950s. They increased the yield of nuclear weapons from a few dozens of kilotons for fission warheads into the hundreds and thousands of kilotons (megatons), and were developed by the United States, Russia, the United Kingdom, France and China.

In the early years of the Cold War, nuclear war threatened destruction in Europe on a scale that would have surpassed that of the Second World War. The arrival of thermonuclear warheads, however, meant that the survival of civilization itself was now seen by many to be at stake. From the 1960s, nuclear-tipped Intercontinental Ballistic Missiles (ICBM), stationed in silos or on submarines, eclipsed the strategic bomber as the main iconic image of nuclear deterrence. Staged, thermonuclear warheads also provided new possibilities to tailor weapons effects, leading to specialized designs, including the 'neutron bomb' which minimized thermal and blast effects in favour of greater initial radiation release.[18]

Since the United States had a significant quantitative and qualitative advantage in nuclear weapons over the Soviet Union during the 1950s, they were central to the strategy of 'massive retaliation'. At that time, US policy saw nuclear munitions as a substitute for expensive conventional forces, and heavily relied on them to deter Communist attacks on its allies – both in Europe and in successive crises in East Asia. Once the Soviet Union achieved the ability to credibly threaten nuclear retaliation against the US homeland, however, reconciling the US pledge to use nuclear weapons in the defence of Europe with the logic of 'mutual assured destruction' became an enduring concern for Western allies.[19] To solve this problem, the United States and NATO turned to a strategy of 'flexible response', which meant that NATO would want to be able to escalate a conflict from the conventional to the nuclear level at a time of its choosing, and that the danger of Soviet retaliation would have to be managed through the ability of NATO to gradually escalate hostilities even during a nuclear conflict.

As such, nuclear strategy as implemented by the United States and Russia shared a number of traits during the Cold War, and continues to do so today, which distinguishes it from the approach to nuclear strategy and deterrence

---

18 By the 1970s, the United States had available ERW that could be fired from standard 155mm artillery pieces. For a history of US nuclear warhead design, see Chuck Hansen, *U.S. Nuclear Weapons: The Secret History* (Arlington, TX: Aerofax, 1988).

19 Lawrence Freedman, *The Evolution of Nuclear Strategy*, 3rd edition (London: Palgrave Macmillan, 2003), 269–314.

by most other nuclear powers, particularly in Asia. Despite significant changes to nuclear strategy over the 40 years of the Cold War, both sides consistently assumed that deterrence had to be based on the credible capability to fight a prolonged nuclear war. To that end, they created nuclear weapons stockpiles in the tens of thousands, and dispersed nuclear warheads onto a wide range of delivery vehicles and weapons systems. A partial explanation for this approach, and a second defining trait of Cold War deterrence, is that the Cold War was at its core a conflict over territorial control of Europe, with both sides perceiving an existential threat with no natural boundaries between them.

In contrast, the strategic situation of most Asian nuclear powers is quite different. Historically, China, India and North Korea were less concerned with deterring existential conventional and nuclear threats, than to prevent other nuclear powers from using their nuclear arsenals as a means to coerce them in crisis situations – as the United States repeatedly did to China in the 1950s. Asian nuclear countries also tend to be separated by significant geographic obstacles from other nuclear powers, which serve to reduce existential threats of invasion. As such, they tend to have a very different approach to nuclear deterrence from that of the major participants of the Cold War. It is generally described as 'minimum deterrence' – the idea that to achieve the strategic objectives of deterrence, it is sufficient to have the credible capability to deliver a small number of nuclear warheads onto the adversary's major cities. The arsenals of China and India in particular have thus remained far smaller than those of the Cold War superpowers: China never held more than 1% of the overall warhead numbers of the United States and Soviet Union during the Cold War, and India did not even weaponize its nuclear capability, demonstrated in the 1974 'peaceful nuclear explosion', until the 1990s.

However, nuclear strategy during the Cold War developed in a way that it was not always necessary for a state to acquire nuclear weapons itself to be able to enjoy the benefits of nuclear deterrence. At the beginning of the Cold War, the United States entered into several close alliance relationships in Europe and Asia, through which nuclear deterrence was 'extended' to its allies. If successful, extended nuclear deterrence reduces the likelihood of aggression against alliance partners, and thus also reduces the incentives for smaller allies to acquire their own nuclear arsenals. Extended deterrence is, however, also more difficult and demanding to sustain than deterrence of threats against oneself, because it requires the ability to escalate a conflict above the level required to protect the smaller ally, in order to deter retaliation against the US homeland. Hence, the credibility of extended deterrence is inseparable from the global nuclear balance, and the potential cost of fulfilling an extended deterrence commitment means that there will often be doubts about the credibility of the posture on the part of the recipient. At the same time, the United States also seeks to avoid entanglement in conflicts involving its allies where its own interests are not vitally engaged.

Managing these conflicting interests of reassurance and entanglement has always been one of the main drivers of US nuclear policy and strategy. During the Cold War, NATO created institutionalized dialogues and integrated planning for nuclear weapons, and the United States provided nuclear warheads for delivery by NATO allies' forces under nuclear 'sharing' arrangements. While both arrangements were implemented in NATO Europe, practical steps to provide credibility for extended deterrence in Asia were limited to the forward-basing of US nuclear weapons, which were withdrawn at the end of the Cold War.[20]

However, in a situation where large nuclear powers were locked in intense competition, or where large nuclear powers have a strong interest in avoiding nuclear escalation of regional conflicts, smaller nuclear powers can also use their own nuclear arsenals to try forcing larger powers to intervene in regional conflicts. During the Cold War, André Beaufre proposed that the French *force de frappe* ultimately strengthened NATO deterrence, because any nuclear exchange between the Soviet Union and France would force the United States to engage with its own nuclear arsenal as well.[21] South Africa thought that by revealing (or using) its small nuclear arsenal in a crisis, it could force Western powers to intervene on its side.[22] Pakistan openly deployed what may have been nuclear warheads at one of its airfields in the 1990 crisis with India, in what is often seen as a (successful) attempt to bring about US diplomatic intervention.[23] What remains a common need, however, of any country acquiring nuclear weapons or contemplating to do so, is to find a credible strategy that explained how, given the overall nuclear global order and regional security situation, the acquisition of nuclear weapons might in practice advance rather than detract from its own security.

The Economic Domain of Nuclear Policy

In parallel with the development of nuclear weapons technology after the Second World War, civilian applications of nuclear fission also captured the imagination of scientists, business and governments. There are several ways in which nuclear fission can be harnessed for economic use – 'economic' being

20 For discussion, see Andrew O'Neil, *Asia, the US and Extended Nuclear Deterrence: Atomic Umbrellas in the Twenty-First Century* (London and New York: Routledge, 2013), chapters 4 and 5.

21 André Beaufre, *An Introduction to Strategy*, trans. Major-General R.H. Barry (New York: Frederick A. Praeger, 1965).

22 See Mitchell Reiss, *Bridled Ambition: Why Countries Constrain Their Nuclear Capabilities* (Washington, DC: Woodrow Wilson Centre Press, 1995), 7–43.

23 Seymour Hersh, 'On the Nuclear Edge', *The New Yorker*, 29 March 1993, available at http://www.newyorker.com/magazine/1993/03/29/on-the-nuclear-edge [accessed 5 August 2013].

used here to encompass all activities that are ultimately contributing to human consumption. 'Peaceful nuclear explosions' were an early interest in the atomic age, including in Australia, but their promise remained unfulfilled. The Soviet Union conducted 124 nuclear explosions until 1988 for geological exploration, oil and gas production, the creation of cavities, channels and dams, extinguishing of natural gas fountains and mining.[24] Today, however, the economic use of nuclear fission is centred almost exclusively on nuclear reactors.

Early prognostications of atomic cars and aeroplanes remained science fiction, and the use of nuclear reactors for civilian propulsion has only found a small niche market in ocean-going icebreakers. Stationary reactors however are applied to a whole range of purposes. Neutron radiation from research reactors has several scientific and commercial applications. As an imaging mechanism, it is used for materials testing, to detect hairline fractures in metal samples such as railroad tracks, for example, in research and development in automotive and other industries, or the biological and chemical sciences. Neutron radiation absorbed by atomic cores can also change one chemical element into another. 'Neutron transmutation doping' makes silicon more conductive for example, and irradiated silicon is used to produce a variety of computer chips, including those found in in mobile phones. 'Neutron activation' produces a variety of specialized isotopes with medical or industrial applications, many of which do not exist in nature. Cobalt-60 for example is used to sterilize food and medical equipment, yttrium-90 for cancer treatment, and molybdenum-99 generates the technetium-99 used in 70,000 medical imaging examinations every day.[25] Other elements are extracted as by-products from the spent fuel of nuclear reactors, including the americium-241 used in household smoke detectors and specialized industrial gauges.

Yet, in terms of their economic importance and prominence, all of these applications are dwarfed by the use of nuclear reactors for energy production. In 2014, 436 reactors in 30 countries produced over 11% of the world's electricity, with a further 71 in various stages of construction.[26] In general, there is considerable difference in countries' reliance on nuclear power. Whereas the

---

24 Pavel Podvig, *Russian Strategic Nuclear Forces* (Cambridge, MA: MIT Press, 2004), 471–80.

25 'Research Reactors', World Nuclear Association, available at http://www.world-nuclear.org/info/Non-Power-Nuclear-Applications/Radioisotopes/Research-Reactors/#.UcPpXvlkPng [accessed 24 June 2013]; Division of Physical and Chemical Sciences, *Research Reactors: Purpose and Future* (Vienna: IAEA, 2010).

26 'Power Reactor Information System', IAEA, available at http://www.iaea.org/PRIS/WorldStatistics/OperationalReactorsByCountry.aspx [accessed 12 August 2014]; 'Nuclear Power in the World Today', World Nuclear Association, available at http://www.world-nuclear.org/info/Current-and-Future-Generation/Nuclear-Power-in-the-World-Today/ [accessed 14 August 2014].

United States is still operating the largest reactor fleet worldwide, France is generating the largest share of electricity from nuclear energy. There are plans for the replacement of a small number of reactors in the United States and United Kingdom, but Germany is phasing out nuclear power, as Italy did during the 1990s. South Korea, Russia, China and India are currently the countries with the largest plans for future construction.[27]

However, these planned projects today remain far smaller, in relative as well as absolute terms, than the past era of nuclear expansion in the 1970s and 1980s. During the 1950s and 1960s, most reactors were relatively small, often experimental and based on a variety of technical principles. Whereas the United States focused on the development of pressurized water reactors (PWR) and boiling water reactors (BWR) using enriched uranium fuel, Canada worked on pressurized heavy water reactors (PHWR), and the United Kingdom on gas-cooled reactors, both of which can operate using natural uranium fuel.

Beginning in the mid-1960s, the use of nuclear power then rapidly increased. Many of the new reactors were much larger, with over 1,000MWe electric output, and most were of either the PWR or BWR type. The nuclear expansion was most dramatic in France, which increased the share of nuclear-generated electricity from 6% in 1970 to 77% in 1995, while tripling overall electricity production. Whereas investment decisions in most countries were taken by local electricity utility companies that chose their own particular reactor designs, France relied on one Westinghouse PWR type, all built by the same company (Framatome) and operated by Electricité de France.[28] Overall, most of the nuclear power capacity in operation today was built during the 1970s and the 1980s. PWR and BWR types dominate with about 61% and 21% respectively of the worldwide commercial reactor fleet. PHWR, mostly of the Canadian CANDU type, represent about 11%, whereas all of the 4% of gas-cooled reactors worldwide are located in the United Kingdom. Soviet-designed RBMK reactors (of the Chernobyl type) represent about 3% of global installed nuclear capacity, and just 0.2% are classified as Fast Breeder Reactors.[29]

Yet, strong expansion in installed reactor capacity during the 1970s and 1980s concealed a dramatic slowing of net new orders in the second half of the 1970s, which was due to a range of reasons that continue to bedevil nuclear power projects to this day. The oil crisis of 1973 reduced economic growth and led to energy efficiency measures that sharply slowed the growth in electricity consumption. The nuclear industry experienced strong growth in construction costs and delays in construction times, and public opinion began to turn against

27 Nuclear Energy Agency, *Nuclear Energy Today* (Paris: OECD, 2012), 8–13.

28 David Bodansky, *Nuclear Energy: Principles, Practices, and Prospects* (New York: Springer, 2004), 18, 33, 41, 48.

29 Nuclear Energy Agency, *Nuclear Energy Today*, 20–22; John Carlson, email to authors, 17 November 2014.

it following the Three Mile Island accident in 1979. Financing also became more difficult in the United States, as the liberalization of the US energy market reduced the long-term pricing power of electrical utility companies. The overall amount of electricity generated from nuclear power however continued to increase because capacity factors – the time that reactors generate electricity and are not shut down for maintenance or refuelling – rose from less than 60% to over 90% by 2002. In recent years, a major focus in many countries with established reactor infrastructure has been to extend the operating life of existing reactors beyond the 40 years for which most were initially licensed.[30]

The relative maturity of the nuclear power generation business is also reflected in the reactor industry structure. While smaller companies remain active in the design and construction of specialized research reactors, the power reactor industry is now largely consolidated into the French Areva, US-Japanese CE-Hitachi and Toshiba-Westinghouse, the Russian Rosatom and Canadian ACEL. With the exception of the latter, these conglomerates include the design and construction of reactors, while South Korean and Chinese companies build foreign designs under licence. Rosatom and, more recently, Areva and Westinghouse are also vertically integrated with activities across the whole nuclear fuel cycle.[31]

From mining over conversion, enrichment and fuel fabrication, each step of the 'front end' of the fuel cycle is increasingly consolidated into a small number of international companies. Canada, Australia and Kazakhstan produce 60% of world uranium, but Cameco, Areva, KazAtomProm and Rio Tinto control 60% of the world's uranium mining and are active in all three of these countries. Rosatom, Areva, Converdyn and Cameco account for 88% of the world uranium conversion market, while Tenex, Eurodif, Urenco and the US Enrichment Corporation account for 95% of commercial uranium enrichment. Areva, Westinghouse, Global Nuclear Fuel and TVEL have a combined share of 84% of fuel fabrication.[32]

Commercial conversion and enrichment of uranium are concentrated in a small number of facilities worldwide that supply the global market. The same is true of commercial reprocessing plants that separate spent fuel to reduce high-level waste and extract plutonium for use in mixed-oxide (MOX) fuels. Only 10% of all reactors globally use MOX fuel. Fast Breeder Reactors that create more (plutonium) fuel than they use have been in operation in a number of countries. Those currently operating FBRs include Russia, India, China and Japan. Others, including FBRs in France that experienced technical difficulties,

---

30 Bodansky, *Nuclear Energy*, 20, 35, 40–41; Nathan Hultman and Jonathan Koomey, 'Three Mile Island: The Driver of US Nuclear Power's Decline?', *Bulletin of the Atomic Scientists* 69, no. 3 (2013): 63–70.

31 Duane Bratt, *Canada, the Provinces and the Global Nuclear Revival* (Montreal: McGill-Queen's University Press, 2012), 71–91.

32 Ibid., 91–2.

have been shut down. Given the size of global uranium reserves and the cost of reprocessing, breeder-based power programs remain uneconomical. Although breeder reactor designs remain under active development as part of future reactor designs,[33] the days of rapid technological development are clearly over.

The Normative Domain of Nuclear Policy

Given the destructive power of nuclear weapons, it was recognized soon after their first use that they were not an unconditional positive for humankind, and that it might be in the best interests of all states to exercise restraint in their development, possession and use. Emerging norms about nuclear weapons could draw on earlier attempts, both practical and philosophical, to limit conflict and the proliferation and use of specific weapons systems. Despite continued attempts and arguments to outlaw nuclear weapons, however, no norm against the possession of nuclear weapons yet exists. Hence, nuclear norms after 1945 resulted in the development of an international nuclear order that William Walker described as 'a managed system of deterrence, and a managed system of abstinence'.[34] Moreover, it is important to note that 'nuclear norms' include the right to the use of civilian nuclear technology, which the NPT recognized as 'inherent'. As such, countries participate in the normative domain of nuclear policy, and the institutions that underpin it, in two related but distinct ways.

As 'norm entrepreneurs', countries can lead or support the political propagation of new nuclear norms, in order to garner international support for the establishment of new modes of behaviour in regards to nuclear weapons.[35] Yet nuclear norms are not uncontested. Indeed, they can be in conflict depending on what proponents seek to avoid – in particular, whether the use (or existence) of nuclear weapons themselves or major war between nuclear powers is framed as the problem that needs to be managed.[36] Even within countries, the discourse about nuclear norms and orders is now arguably conducted between different 'epistemic communities', which disagree not only about aims and objectives but even interpretations of facts.[37] Through participation in such discourse, through political advocacy and international propagation of particular nuclear norms, governments and civil society are able to participate in an important way in the normative domain of nuclear policy formation.

---

33 Nuclear Energy Agency, *Nuclear Energy Today*, 35; Bodansky, *Nuclear Energy*, 189.

34 William Walker, 'Nuclear Order and Disorder', *International Affairs* 76, no. 4 (2000): 703.

35 See Finnemore and Sikkink, 'International Norm Dynamics and Political Change'.

36 See Lawrence Freedman, 'Disarmament and Other Nuclear Norms', *Washington Quarterly* 36, no. 2 (2013): 93–108.

37 Joachim Krause, 'Enlightenment and Nuclear Order', *International Affairs* 83, no. 3 (2007): 483–99.

In addition, countries can also participate in the normative domain through the development and implementation of international regulatory measures that implement and verify states' commitments. Attempts to establish and embed norms regarding the possession and use of nuclear technology must be reconciled in practice with its dual-use nature and the perceived promise of nuclear technology for economic development. This creates an additional layer of political and practical complication that has given rise to a wide range of international and domestic institutions over time. Managing the risks of nuclear proliferation, safety and security, while preserving the right to legitimate civilian uses of nuclear technology, requires a raft of domestic and international regulatory measures. Safeguarding nuclear facilities and materials to verify NPT commitments in particular remains an enduring challenge for the international community, especially as technology continues to develop. Over time, a range of approaches has emerged that comprise material accountancy, seals and physical containment measures, environmental sampling, export controls and means to intercept international shipments of nuclear materials.

Normative frameworks for the strategic and civilian use of nuclear technology find expression in a range of different international agreements and institutions. Nuclear arms control agreements emerged during the Cold War between the United States and Soviet Union to strengthen and stabilize a situation of mutual deterrence. Besides major treaties to limit (SALT treaties), reduce (START treaties) or eliminate (INF treaty) certain types of nuclear delivery systems, they comprised a range of agreements to reduce the likelihood of inadvertent conflict. Most of these agreements, however, remained bilateral, and during the Cold War verification relied on the 'national technical means' of both sides – in particular, optical reconnaissance and electronic intelligence satellites.

A range of international agreements codify commitments to nuclear non-proliferation and nuclear abstinence. Most prominent among these is the NPT itself, which is arguably the linchpin of nuclear order today. Verification of NPT commitments rests on nuclear safeguards on nuclear materials and installations under the NPT as well as bilateral agreements, and which are mostly administered by the IAEA. To this end, member states must create a national State System of Accounting for and Control of Nuclear Material (SSAC) in national law, whose work is in turn verified by IAEA personnel. Non-proliferation efforts are also supported by domestic systems of export controls, which industrialized countries developed since the 1970s to check other countries' ability to acquire nuclear technology and related components. Export control efforts are supported by international groups, especially the Nuclear Suppliers Group, which maintain lists of controlled items and exchange information about best practice, as well as about attempts to circumvent controls. Since 2003, countries participating in the Proliferation Security Initiative are also strengthening their ability to intercept suspect shipments en route to countries of concern.

In addition, a range of nuclear weapons-free zones exist around the globe, including in Southeast Asia and the South Pacific. An important agreement not yet in force is the CTBT, which built on the earlier limited test ban treaty between the United States and Soviet Union. While it is not yet in force, the CTBT is already verified by a network of international monitoring stations administered by member states (including through Geoscience Australia).

Finally, normative frameworks have also developed in regards to dangers other than the use of nuclear technology for military purposes. In particular, these relate to the way in which countries should balance the civilian use of nuclear technology with concerns about radiation exposure to the public and the environment, be it through accidents (nuclear safety) or through theft and deliberate release as part of terrorism and other criminal activity (nuclear security).[38] In the United States, for example, the US Atomic Energy Commission was split in 1974 into the US Energy Research and Development Administration on the one hand, and the US Nuclear Regulatory Commission on the other hand.[39] Article 8(2) of the 1994 Convention on Nuclear Safety requires that states ensure an effective separation between the regulatory body responsible for maintaining a regulatory framework to govern the safety of nuclear installations (licensing, inspection, enforcement), and organizations responsible for promotion or utilization of nuclear energy.[40] Hence, there is now a considerable body of model domestic legislation that countries are expected to adopt as best practice to manage the use of civilian nuclear technology.[41] Normative aspects of nuclear policy thus extend from high-level international diplomacy, to the detailed technical, legal and governance aspects within each country involved with nuclear technology.

Unifying Elements of Nuclear Policy: The State, Security and Technological Expertise

The strategic, economic and normative domains involve very different objectives, actors and even philosophical approaches. Nonetheless, it makes sense to think of all three as part of a coherent field of nuclear (fission) policy –

38 The authors would like to acknowledge useful comments and suggestions from Kalman Robertson on this section.

39 See J. Samuel Walker and Thomas R. Wellock, *A Short History of Nuclear Regulation, 1946–2009* (Washington, DC: US Nuclear Regulatory Commission, 2010).

40 International Atomic Energy Agency, *Convention on Nuclear Safety*, INFCIRC/440 (Vienna: IAEA, 1994), available at http://www.iaea.org/Publications/Documents/Infcircs/Others/inf449.shtml [accessed 20 July 2014].

41 See for example Carlton Stoiber, Abdelmadjid Cherf, Wolfram Tonhauser and Maria de Lourdes Vez Carmona, *Handbook on Nuclear Law: Implementing Legislation* (Vienna: IAEA, 2010).

even if that coherence is not necessarily recognized in practice, or reflected in the framing of bureaucratic or political approaches to specific issues. The coherence of nuclear policy ultimately derives from the position of the state as the central, if not sole actor in each domain; from the fact that outcomes in one domain – in terms of national capabilities and policy positions – are inseparable from outcomes in the others; and from the importance of nuclear technological expertise for activities across all aspects of nuclear policy.

For obvious reasons, strategic applications of nuclear technology are the almost exclusive domain of the state – related as they are to military capabilities and national defence policy. Private industry has a role in the development and maintenance of nuclear weapons in several nuclear weapons states, and defence policy is made in the context of national debates in which think tanks, media, political parties and academia also participate. In the end, however, decisions about nuclear weapons capabilities, or participation in extended nuclear deterrence relationships, remain fully the purview of government. The same is also true for participation in international arms control and non-proliferation agreements, and the related establishment of a national regulatory framework for export controls and nuclear materials accounting.

The role of the state is more indirect, but nonetheless central to economic applications of nuclear technology as well, despite the commercial nature of most of the nuclear industry today. The construction and operation of nuclear reactors requires a permissive legislative framework, and states regulate nuclear industry in relation to environmental, safety, security, non-proliferation and occupational health and safety concerns. Again, the centralized approach of France stands out in this regard, whereas opponents of nuclear energy tend to have far more avenues to stop new projects in more federalized systems of government, including Australia.[42] One of the most extreme, and costly, examples of such regulation is the US reactor at Shoreham, Long Island, which only briefly operated in 1985 but was then dismantled over local authorities' refusal to approve evacuation plans.[43]

Some nuclear technology companies remain partially or wholly in state ownership, even though they are generally thoroughly commercial in outlook. Almost all reactor designs in use today originated in state-led research institutes, even if they were then later developed further by industry. Nonetheless, the role of the state remains central even to the development of nuclear reactor technology. In the 'Generation IV' initiative that began in 2001, Argentina, Brazil, Canada, China, EURATOM, France, Japan, South Korea, Russia, South Africa, Switzerland, the United Kingdom and the United States cooperate, through their national research institutes, in the development of future reactor designs.

---

42 Bodansky, *Nuclear Energy*, 48.

43 Joan Aron, *Licensed to Kill? The Nuclear Regulatory Commission and the Shoreham Power Plant* (Pittsburgh, PA: Pittsburgh University Press, 1998).

The aim of participating countries is to develop new technologies and concepts for reactors with improved safety, sustainability and proliferation resistance.[44]

Governments are also central to the business case of nuclear power reactor projects. About 60% of the total lifecycle cost of a nuclear power reactor is incurred during construction, before it produces any electricity for sale. Fuel only makes up 10–15% of the total cost, half of which is attributable to the uranium itself. The cost of finance, possible delays in construction and certification, and the volatility of long-term electricity prices are thus critical risks for the economic viability of new nuclear reactors. In liberalized electricity markets, the construction of new or replacement reactors thus generally requires government guarantees, or other help with finance, to share the financial risk involved.[45]

In the production of medical isotopes, governments have had an even more direct influence on the industry. While many processing steps are now in the hands of privatized companies, the irradiation itself continues to rely on government-funded research reactors. Traditionally, irradiation prices have reflected marginal costs, excluding the depreciation of the reactor, and there is generally no provision in contracts for the maintenance of spare capacity (because of its short half-life, molybdenum-99 must be continuously produced). As a result, the current market structure for medical isotopes continues to rely on substantial, if indirect government subsidies.[46]

In terms of outcomes, states' use of nuclear weapons and reliance on nuclear deterrence on the one hand, and their participation in normative frameworks that place limits on their own as well as other's strategic applications of nuclear technology on the other hand, are ultimately both motivated by a quest for security. Differences lie in the source of insecurity that countries seek to manage: The purpose of nuclear weapons and deterrence is to manage aggression by other countries' existing capabilities, whereas non-proliferation seeks to forestall future threats by limiting access to nuclear technology and avoiding unnecessary security dilemmas between non-nuclear states. The aim of arms control is to maximize stability between countries in conflict, while those arguing for nuclear disarmament mostly fear nuclear weapons and the danger of their inadvertent or deliberate use itself. The ultimate objective of security, however, is the same, even if each approach can lead to conflicting priorities and conclusions about the extent to which a country should make strategic use of nuclear technology.

---

44 Bodansky, *Nuclear Energy*, 470–72; Nuclear Energy Agency, *GIF Symposium Proceedings: 2012 Annual Report* (Paris: OECD, 2012), esp. 9–12.

45 Financing, assuming a 5% interest rate, is included in the construction cost. Nuclear Energy Agency, *Nuclear Energy Today*, 80–81; Mycle Schneider and Antony Froggatt, *The World Nuclear Industry Status Report 2012* (Paris: Mycle Schneider Consulting, 2012), 32–40.

46 Nuclear Energy Agency, *The Path to a Reliable Supply of Medical Radioisotopes: Nuclear Energy in Perspective* (Paris: OECD, 2011).

Bridging the economic and strategic domains, the use of nuclear energy for power production can have implications for security as well. Nuclear power plants only require about 25 tons of fuel per year, which can be stored several years in advance.[47] Hence, nuclear energy provides a source of power that does not rely on continuous bulk shipments of fuel from overseas, and whose price is not correlated with that of conventional fuel sources. In general, security of supply considerations have thus been central to the decision of all countries which have made large use of nuclear energy for power production, and to decisions to develop breeder reactor technology in particular.[48] And insofar as countries need to import uranium for their reactors, long-term relationships with supplier countries can then also become highly relevant for their energy security.

Moreover, economic applications of nuclear technology also have strategic implications insofar as they would make it possible for a country to more quickly, cheaply, or more reliably develop nuclear weapons than it would be able to do otherwise, should it decide to do so. The concept of nuclear 'latency' captures this largely inevitable and usually inadvertent side-effect of nuclear facilities, fissile material and expertise developed for purely commercial or economic reasons. In a more sinister manner, countries are 'hedging' if an increase of nuclear infrastructure and expertise is deliberately sought in order to increase nuclear latency – a policy that is arguably incompatible with legitimate civilian use under the NPT, but can be difficult if not impossible to discern from the outside.[49]

Attempts to determine the extent of nuclear latency are difficult, and often focus primarily on facilities and stocks of fissile materials over less easily assessed technological expertise.[50] Developing new technologies and industry structures that limit latency has been an important practical challenge for nuclear policy throughout the decades. Of particular importance here are facilities that can produce fissile material, including uranium ore processing and milling capability, conversion capability, enrichment capability, fuel fabrication capability (including zirconium alloy processing capability), heavy water production capabilities and spent fuel processing facilities.

---

47 In contrast, a comparable coal plant would require about 3 million tons of coal per year. See Nuclear Energy Agency, *Nuclear Energy Today*, 88.

48 While resource-poor Japan and France are the most prominent examples, the same is also true for India and its long-term plans for the development of a thorium-based fuel cycle dating back to the 1950s.

49 See John Carlson's speech as part of the Australian Ambassador's 2012 Speaker Series: John Carlson, 'The "Prague Agenda" and Nuclear Energy' (Australian Embassy, Washington, DC, 13 November 2012), available at http://www.lowyinstitute.org/publications/prague-agenda-and-nuclear-energy [accessed 20 May 2014].

50 Scott D. Sagan, 'Nuclear Latency and Nuclear Proliferation', in William Potter, ed., *Forecasting Nuclear Proliferation in the 21st Century* (Stanford, CA: Stanford University Press, 2010), 80–101.

In general, experience of a country in the design, development, building and operation of nuclear weapons, of nuclear reactors, or of safeguards and other non-proliferation mechanisms, is neither necessary nor sufficient for proficiency in the other areas. Nonetheless, thorough understanding of nuclear technology is ultimately necessary for activity in any of these areas, and the resulting expertise would be of use if a country wanted to expand its nuclear activities into other activities.[51] Nuclear expertise usually resides in a range of institutions and individuals, including nuclear physics and engineering departments at universities, states' nuclear technology research organizations, defence research agencies and a possible nuclear weapons complex, as well as private companies that are involved in the design, construction and operation of nuclear reactors and related technologies. It includes nuclear engineers, nuclear physicists and radiochemists, as well as mechanical engineers and metallurgists.[52]

For economic applications, a country must draw from these experts for its research and development personnel, designers, lab technicians, reactor operators and reactor safety specialists, as well as policy makers and regulators. For strategic application, a country requires nuclear weapons designers and engineers, as well as policy makers and analysts to assess other countries' capabilities. And even for participation in non-proliferation, nuclear safeguards and export control regimes, expertise in nuclear technology is essential – particularly if a country wants to actively participate in the implementation of international safeguards through the provision of inspectors, analysis of samples, or the development of new technologies for safeguards. Although the IAEA implements international safeguards on nuclear facilities, research and development of safeguards technology is largely based on individual countries' technical support thorough Member State Safeguard Support Program agreements, including one signed by Australia in 1980.[53]

Given the specialized nature and long service life of many nuclear installations, knowledge management over time is a challenge even for commercial plant operators.[54] Governments as well cannot generally rely on the civilian sector alone to provide sufficient employment prospects and education programs to generate and maintain the nuclear expertise they require to engage in the technical

51 On this theme, see Matthew Fuhrmann, *Atomic Assistance: How 'Atoms for Peace' Programs Cause Nuclear Insecurity* (Ithaca, NY: Cornell University Press, 2012).

52 The authors would like to acknowledge useful comments and suggestions from Kalman Robertson on this section.

53 H. Kurihara, 'Role of Support Programmes in Safeguards', *IAEA Bulletin* 25, no. 1 (1988): 13–17. See also Department of Safeguards, *Development and Implementation Support Programme for Nuclear Verification 2012–2013*, STR-371 (Vienna: IAEA, 2011).

54 International Atomic Energy Agency, Nuclear Knowledge Management Section, *The Impact of Knowledge Management Practices on NPP Organizational Performance – Results of a Global Survey*, IAEA-TECDOC-1711 (Vienna: IAEA, 2013).

aspects of all three domains of nuclear policy. Instead, public funding and some measure of coordination of research and education activities are required to develop and maintain a national nuclear technological base.[55] This is particularly so as many nuclear facilities, and a significant sector of many countries' nuclear workforce, date to the expansion of nuclear power production during the 1960s. Hence, the need to preserve and develop nuclear expertise as the basis for nuclear policy making, and for the application of nuclear technology in general, is common to all countries interested in managing the challenges arising from the dual use nature of nuclear fission – including in Australia.[56]

Conclusion

In this chapter, we have outlined the various elements of nuclear policy and argued that integrating these elements into an overall perspective is useful to thinking about how countries deal with nuclear questions. Issues that pertain to economic profit in the nuclear realm inevitably raise questions about strategic risk regarding proliferation that in turn provoke concerns about the durability of norms with respect to responsible nuclear behaviour internationally. This holistic approach to understanding nuclear policy is highly relevant to Australia's circumstances as a major exporter of nuclear fuel, as a state that has persistent concerns about the future proliferation environment in Asia, and as a country that prides itself as a key 'norm entrepreneur' in promoting new initiatives on nuclear issues. But before we can begin to appreciate how these factors intersect in the contemporary and more recent contexts, it is necessary to examine Australia's nuclear options, and how governments made significant nuclear choices, in the period between 1945 and the early 1970s.

55 Anil Kakodkar, 'Managing Nuclear Knowledge: A Governmental Perspective', *International Journal of Nuclear Knowledge Management* 1, no. 1/2 (2004): 33–8.

56 For a review of global efforts and recommendations, see Nuclear Energy Agency, *Nuclear Education and Training: From Concern to Capability* (Paris: OECD, 2012).

Chapter 2

Exploring Australia's Nuclear Options: Second World War to the Whitlam Government (1945–75)

Large uranium ore deposits across northern and southern Australia were initially discovered in the late 19th century. But not until the discovery of atomic fission – and its potential for industrial and military application – did Australian policy makers realize that the mineral constituted a strategic resource with strategic implications and potential economic value. Even prior to the first test of an atomic device in New Mexico in 1945, Australia was part of the nuclear age. An Australian physicist, Mark Oliphant, was closely involved in research that led to the development of the first nuclear weapon, and known Australian uranium deposits factored into the Anglo-American atomic collaboration agreement signed in Quebec in 1943.[1] This reflected and reinforced a broader interest in nuclear issues on the part of Australian policy makers and the potential they perceived in the emerging nuclear fuel cycle to contribute to Australia's postwar national development. While Australia was kept at arm's length as the Manhattan Project evolved between 1942 and 1945, the nuclear attacks on Hiroshima and Nagasaki – and the evident potential of these weapons to inflict even greater damage – had a major impact on Australian policy makers.

Australia's Weapons Choice

Australia's capacity to manufacture nuclear weapons was widely acknowledged in much of the early Western literature on nuclear proliferation. As a relatively advanced industrial state possessing a healthy scientific base and noteworthy deposits of uranium ore, Australia was often identified as a likely candidate for nuclear weapons acquisition throughout the 1950s and 1960s.[2] As Jim

1 Stewart Cockburn and David Ellyard, *Oliphant: The Life and Times of Sir Mark Oliphant* (Adelaide: Axiom Books, 1981); Jonathan Helmreich, *Gathering Rare Ores: The Diplomacy of Uranium Acquisition, 1943–1954* (Princeton, NJ: Princeton University Press, 1986), chapter 1.

2 See for example Howard Simons, 'World-Wide Capabilities for Production and Control of Nuclear Weapons', *Daedalus* 88, no. 3 (1959): 395; Leonard Beaton, 'Nuclear Fuel-For-All', *Foreign Affairs* 45, no. 4 (1967): 662.

Walsh has documented, serious consideration by Australian governments of a national nuclear weapons option occurred between 1956 and 1961, when the Menzies government explored the possibility of directly procuring nuclear weapons from Britain, and between 1964 and 1972 when successive Coalition governments contemplated acquiring nuclear weapons through a national effort.[3] This second phase spanned a time during which Australia's external security environment became increasingly uncertain. Following the Menzies government's involvement in 'Confrontation' with Indonesia in the early 1960s, and China's explosion of a nuclear device in 1964, Australia was faced with what appeared to be a deteriorating regional security climate when Britain (in 1967) and the United States (in 1969) flagged a substantial scaling down of their respective strategic presence in the Asia-Pacific. With looming defeat in Vietnam, it became clear that Australia's postwar strategic doctrine of 'forward defence' was increasingly untenable and that defence policy would somehow need to be reconfigured to adjust to these circumstances. As Hedley Bull was to observe subsequently, Australia's changing external security situation alone presented policy makers with what seemed to be a compelling motive to acquire nuclear weapons.[4] Why, then, did Australia not opt for a sovereign nuclear weapons capability?

This question can best be addressed by focusing on the internal dynamics of Australian nuclear policy making in the late 1950s and 1960s, and the changing international context of the late 1960s. Although developing a nuclear force would have entailed significant financial outlays, there can be little doubt that Australia had the latent technological capability to 'go nuclear' at some point in the future if it so decided during these years. While expensive, a successful weapons program would have been technically feasible.[5] It is clear that *motive* was the independent variable determining the approach of successive Coalition governments to the question of nuclear acquisition throughout the 1950s and 1960s. Two points immediately stand out when surveying this period.

The first is that at no stage did any government give official countenance for work to proceed on an Australian nuclear weapons program. Although periodically attracted by this option, foreign policy elites also displayed a high degree of caution when deliberating nuclear issues. However, successive

---

3 Walsh, 'Surprise Down Under: The Secret History of Australia's Nuclear Ambitions'.

4 Hedley Bull, 'Australia and the Nuclear Problem: Some Concluding Comments', in R. O'Neill, ed., *The Strategic Nuclear Balance: An Australian Perspective* (Canberra: Australian National University Press, 1975), 139–40.

5 For discussion, see Thomas B. Millar, *Australia's Defence*, 2nd edition (Carlton: Melbourne University Press, 1969), 157–68; A.L. Burns, 'Australia and the Nuclear Balance', in Harry Gelber, ed., *Problems of Australian Defence* (Melbourne: Oxford University Press, 1970), 150–51; Ian Bellany, *Australia in the Nuclear Age: National Defence and National Development* (Sydney: Sydney University Press, 1972), 61–80.

governments were careful not to rule out the possibility of nuclear acquisition either. Although having previously registered its opposition to any further spread of nuclear weapons,[6] in 1962 the Menzies government refused to provide the then Acting Secretary-General of the United Nations with an assurance that Australia would refrain from acquiring nuclear weapons following a request for such a declaration, noting that 'nations must be free to look after their own security in accordance with Article 51 of the [UN] Charter'.[7] The key theme underlying this period was that foreign policy elites consciously and consistently chose to hedge their bets in relation to the nuclear weapons option. This accords with Michael Carr's broader observation that Australian government attitudes to nuclear issues in the postwar period were generally based on the principle of 'keep[ing] open as many options as possible'.[8] Indeed, the Menzies and Gorton governments in particular sought to preserve Australia's option for acquiring nuclear weapons, yet at the same time avoided approving research and development programs directly related to weapons development.

Importantly, Western defence strategy against conventional attack during the 1950s and early 1960s was based on 'massive retaliation', including the first and wide-scale use of both tactical and strategic nuclear weapons to support military operations.[9] The 1956 *Strategic Basis of Australian Defence Policy* stated that

> Planning can reasonably proceed on the basis that Australian forces engaged in operations in conjunction with United Kingdom and United States forces, in accordance with common treaty obligations, will be supported by nuclear action by the United Kingdom and United States, when circumstances require such support. This is inherent in the SEATO strategic concepts developed to date for the defence of South East Asia in limited war, and can be expected similarly to apply in other cases as plans are developed to meet other situations (e.g. global war and contingency plans for the defence of Malaya). The form of support might well be by making available for service with our forces elements armed with and capable of using nuclear weapons or by making such weapons available to our forces *in the field* under certain operational circumstances.[10]

6 See Menzies' comments in *Commonwealth Parliamentary Debates* (Representatives), 19 September 1957, 798. (Hereafter cited as *C.P.D.*)

7 'Letter sent by the Minister for External Affairs to the Acting Secretary-General of the United Nations concerning Resolution 1664 (XVI), 15 March 1962', tabled in *C.P.D.* (Representatives), 5 April 1962, 1381.

8 M. Carr, 'Australia and the Nuclear Question: A Survey of Government Attitudes, 1945 to 1975' (unpublished Master of Arts Thesis, University of New South Wales, 1979), 137.

9 For discussion of the rationale of 'massive retaliation' in President Eisenhower's thinking see Ira Chernus, *Apocalypse Management: Eisenhower and the Discourse of National Insecurity* (Stanford, CA: Stanford University Press, 2008), 64–5.

10 Defence Committee, *Strategic Basis of Australian Defence Policy* (Canberra: Australian Department of Defence, 11 October 1956), para. 22. Emphasis in original.

The question was thus not whether Australia should plan on the basis of a nuclear defence, but whether it could and should seek more formalized nuclear sharing arrangements, as they were then practiced in NATO. In 1956 the Defence Committee of Cabinet recommended to senior government Ministers that Australia initiate discussions with Britain over the receipt of tactical nuclear weapons for use by Australian armed forces. Between 1956 and 1958, Australian and British officials engaged in tentative consultations over the feasibility of a bilateral transfer arrangement, with key sections of the British defence establishment expressing a willingness to respond favourably to any future transfer request from Australia.[11]

At this time some Australian officials also began to express doubts about the credibility of US extended deterrence guarantees in the event of a global or regional conflict, prompting suggestions that Canberra may need to consider the development of an indigenous nuclear option. In August 1956, for example, Malcolm Booker, head of the South and South East Asia Branch of the Department of External Affairs, outlined to the Secretary of the Department, Sir Arthur Tange, a line of argument that would recur throughout the coming decades, and that highlights the preeminent strategic anxieties of Australian policy makers at the time. '*In the long term*', Booker argued, 'Australia cannot rely on the United States being able to protect us with either conventional or thermo-nuclear weapons', and that the consequences of this for Australia 'will clearly be grave in the extreme'.[12] For Booker Australia, in the absence of US nuclear or conventional deterrence guarantees, would be 'powerless to prevent the expansion of the Communists in Asia, or check another southward drive by (a possibly Communist) Japan. Nor would we be able … to protect the mainland of Australia'.[13] Such considerations meant that Australia must 'review fundamentally and realistically the question of whether we can in the longer term rely on our alliances for our protection or whether we must provide for it ourselves' through either the 'build-up of conventional forces (including atomic weapons)' or the development of 'our own thermo-nuclear capacity'.[14]

In 1957 the Menzies government also participated in secret negotiations with American officials regarding the possibility of obtaining tactical nuclear weapons from the United States.[15] However, in spite of these bilateral approaches

11 See 'Procurement of nuclear weapons for Australian forces', NAA A1945, 186/5/3; Walsh, 'Surprise Down Under: The Secret History of Australia's Nuclear Ambitions', 3–5.

12 'Thermo-Nuclear Isolationism', Minute from Booker to Tange, 6 August 1956, in Wayne Reynolds and David Lee, eds, *Australia and the Nuclear Non-Proliferation Treaty, 1945–1974* (Canberra: Department of Foreign Affairs and Trade, 2013), 30. Emphasis in original.

13 Ibid.

14 Ibid.

15 M. Taylor, 'The Day We Tried to Get the Bomb', *The Australian*, 1 January 1988.

and the doubts regarding US extended deterrence expressed by officials such as Booker, the Coalition government exhibited caution when confronted with concrete proposals for Australia developing its own nuclear arsenal. Menzies himself was also far less energetic in his approaches to Britain on the issue than some of his civilian and military advisors would have preferred.[16] In the end, the view that nuclear weapons are best left in the hands of the great powers, and that Australia's security would ultimately be best served through the restraint of smaller countries underpinning that nuclear order, prevailed.[17]

By the mid-1960s, doubts about the credibility of extended deterrence increased among many US allies as the Soviet Union gained the capability to retaliate against the US homeland, and the United States moved to a strategy of 'flexible response' that de-emphasized the early use of nuclear weapons.[18] In Australia, the Chinese entry to the nuclear weapons club, doubts about the depth of US and UK commitment to Southeast Asia, and the prospect of conflict with a belligerent Indonesia under Sukarno reinforced these fears and strengthened arguments for a sovereign nuclear weapons capability as a substitute for, rather than complement to, an alliance-based defence posture.

But even during the height of official sympathy for a sovereign nuclear option (under the Gorton government), Australian policy makers remained formally circumspect concerning their nuclear ambitions. Gorton's approval in 1969 of the construction of a Commonwealth nuclear power reactor at Jervis Bay was interpreted in some quarters as indicative of the Coalition government's interest in an Australian nuclear weapons program.[19] The project's tender specifications that 'the reactor must be capable of using indigenous nuclear fuel' (i.e. uranium)[20] were construed by some as a transparent attempt to avoid any potential restrictions on a prospective nuclear weapons program by ensuring that the Jervis Bay reactor could yield weapons-grade material independent of external safeguarded fuel supplies.[21] However, a mere four months after the Jervis Bay announcement, the government revealed that, notwithstanding its obvious qualms over the Treaty, it would sign the NPT. By attaching several

---

16 See Walsh, 'Surprise Down Under: The Secret History of Australia's Nuclear Ambitions'.

17 See Leah and Lyon, 'Three Visions of the Bomb', on the 'Menzian' view.

18 For a critical reexamination of 'flexible response' see Francis J. Gavin, 'The Myth of Flexible Response: United States Strategy in Europe during the 1960s', *International History Review* 23, no. 3 (2001): 847–75.

19 Ian Bellany, 'How Nuclear Should Australia Go?', *The Bulletin*, 23 August 1969, 43–5.

20 Australian Atomic Energy Commission. 'Jervis Bay Nuclear Power Project', *Atomic Energy in Australia* 13, no. 2 (April 1970): 32; Hymans, 'Isotopes and Identity: Australia and the Nuclear Weapons Option, 1949–1999', 10–11.

21 See for instance S. Encel and A. McKnight, 'Bombs, Power Stations, and Proliferation', *Australian Quarterly* 42, no. 1 (March 1970): 17–21.

conditions to its signature, Australia fell into the category of those states which Holst noted at the time 'opted for a sort of provisional virginity making their nuclear abstention dependent on the non-occurrence of events which would force them to reconsider'.[22] Signing the NPT meant that Australia saw no immediate need to acquire nuclear weapons – but it did not mean that it had decided to completely forego the option to do so in the future.

The second salient point about the two phases in question is that at no stage did those bureaucratic actors in favour of Australia acquiring nuclear weapons gain predominance in the policy making process. The most entrenched bureaucratic advocates of an Australian nuclear force were located in the senior ranks of the AAEC, elements of the Department for Supply, sections of the defence establishment – particularly the Royal Australian Air Force (RAAF) – and among a small, though vocal, number of Parliamentary backbenchers. Yet, at no point were advocates successful in persuading senior government policy makers of the merits of Australia acquiring nuclear weapons; the notion of an Australian nuclear force never received anything approaching majority support within Coalition ranks, let alone within Cabinet.

In the late 1950s, the RAAF had been a keen supporter of nuclear weapons procurement, and Air Chief Marshal Frederick Scherger was one of the more outspoken supporters of acquisition after his retirement.[23] Following its creation in 1953, the AAEC remained the most consistent bureaucratic supporter of nuclear acquisition. Its chairman after 1957, Philip Baxter, sought to persuade government foreign policy elites of the merits of commencing preparations for an Australian nuclear weapons program. As early as 1958, for instance, Baxter proposed constructing a nuclear facility at Mount Isa capable of producing plutonium that could be stockpiled for use in the manufacture of nuclear weapons.[24] During what is generally regarded as the peak of his bureaucratic influence as chairman of the AAEC in the late 1960s, Baxter continued to advocate the nuclear option; his opposition to Australia acceding to the NPT derived in no small part from a view that membership of the Treaty would severely circumscribe Australia's future nuclear progress in both the military and economic domains.[25] These views were shared by the then Minister for

22 J.J. Holst, 'Small Powers in a Nuclear World', in A. Schou and A.O. Brundtland, eds, *Small States in International Relations* (Stockholm: Almqvist and Wiksell, 1971), 193.

23 'Scherger Wants Us to Have Nuclear Bombs', *The Australian*, 8 October 1971.

24 Cawte, *Atomic Australia*, 108.

25 A 1968 article published under a pseudonym (but widely believed to be the work of Baxter) was outspokenly critical of the NPT, particularly in relation to the inspection provisions under the Treaty aimed at verifying states' compliance with their declared non-nuclear status. See X, 'Australian Doubts on the Treaty', *Quadrant* 12, no. 3 (May/June 1968): 30–34. It is worth noting that Baxter was considered by some as the 'prime adviser' to the Gorton government on the NPT after 1968. B. Martin, *Nuclear Knights* (Canberra: Rupert Public Interest Movement, 1980), 42.

National Development, David Fairbairn, who in 1969 publicly expressed scepticism towards the NPT and the supposedly deleterious impact it would have on Australia's ability to 'go nuclear' if any government chose that path.[26]

Why didn't these pro-acquisition sentiments gain ascendancy under Coalition governments in the late 1950s and 1960s? A major factor why they did not was the attitude of Menzies himself. Despite his government's discussions with Britain over weapons procurement, Menzies remained lukewarm about suggestions that Australia acquire nuclear weapons and displayed little interest in proposals which he regarded as having the potential to undermine the status quo by altering Australia's supposed reliance on its 'great and powerful friends' in the event of any mortal challenge to national security.[27] Prime Minister John Gorton was notably more receptive to these pro-acquisition arguments. Unlike Menzies, however, Gorton faced a far less favourable climate in which to actively push for an Australian nuclear capability, had he been so inclined (moreover, Gorton came to power after Sukarno's regime in Indonesia had lost power to the pro-Western Suharto). Growing international support for the NPT after its conclusion in 1968 placed many developed countries with latent nuclear capabilities in a quandary: they could declare their support for the Treaty and publicly disavow the weapons option, or refuse to support it and remain subject to widespread suspicion as to their nuclear intent. Thus, when Australia enjoyed what seemed to be fairly broad scope to procure nuclear weapons during the late 1950s, it had a Prime Minister who was ultimately disinclined towards the weapons option. When it finally had a Prime Minister who *was* inclined towards that option, Australia clearly no longer had the latitude for nuclear acquisition it had enjoyed during previous decades.

Moreover, unlike Menzies, Gorton's personal authority on issues of foreign policy remained tenuous during his time as Prime Minister.[28] This was crucial in limiting his influence on nuclear policy and permitted those bureaucratic actors opposed to a nuclear weapons program to exercise influence in ensuring that Australia remained a non-nuclear state. The most powerful of these, the Departments of Defence, External Affairs and Treasury, argued that the excessive cost of any weapons program and the potential strain it could impose on bilateral ties with the United States outweighed any strategic benefits that could be accrued as a result of Australia acquiring nuclear weapons.[29]

---

26 'Don't Close Door to A-Weapons: Minister', *The Sydney Morning Herald*, 14 July 1969.

27 Walsh, 'Surprise Down Under: The Secret History of Australia's Nuclear Ambitions', 3.

28 Russell Trood, 'Prime Ministers and Foreign Policy', Patrick Weller, ed., *Menzies to Keating: The Development of the Australian Prime Ministership* (Carlton: Melbourne University Press, 1992), 167.

29 Cawte, *Atomic Australia*, 117; Carr, 'Australia and the Nuclear Question: A Survey of Government Attitudes, 1945 to 1975', 144; Christopher Hubbard, 'From

With senior Cabinet ministers ambivalent over nuclear acquisition, and with his own political judgment under scrutiny following the 1969 election, Gorton simply did not have the authority to 'carry' any decision to proceed with an Australian nuclear weapons program.

On a broader level, it is also worth emphasizing the splintered nature of pro-acquisition forces in Australia during the late 1950s and 1960s. A key ingredient in those states that determine to proceed with a weapons program is the successful creation of coalitions among pro-nuclear individuals and groups.[30] This ingredient did not exist in the Australian case. At the bureaucratic level, pro-acquisition forces appeared fragmented and, to employ Allison's terminology, 'semi-feudal, loosely allied [and] each with a substantial life of its own'.[31] Though it would be inaccurate to conclude that those in favour of nuclear acquisition enjoyed no influence among policy makers – in Gorton's case they clearly did – at no stage were they powerful enough to form coalitions to foster broader political support for their ideas at the national level.

Nuclear Power and National Development

Flirtation with the option of acquiring nuclear weapons would remain a persistent trend in Australian official discussion until the early 1970s. But while successive governments were uneasy about Australia's strategic circumstances in the 1950s and 1960s, consideration of a weapons option was merely one dimension of Australian nuclear policy at this time. The other important dimension was the perceived role that nuclear power could play in driving forward Australia's national economic development strategy and promoting the country's profile as a player of influence in global nuclear diplomacy. In the postwar period, the Chifley and Menzies governments took very seriously the idea that the nuclear fuel cycle could yield substantial benefits for Australia in the economic and foreign policy domains, and this was reflected in the creation of a domestic nuclear infrastructure and key appointments to ensure Australia maintained a strong engagement with the nuclear fuel cycle and was recognized as a 'player' in international nuclear affairs.

As early as 1946, Prime Minister Chifley envisaged that nuclear energy would have an enduring economic impact and that Australia should actively prepare to exploit its benefits in the coming decades. In particular, he identified major uranium mining and extraction, ambitious scientific research and development in

---

Ambivalence to Influence: Australia and the Negotiation of the 1968 Nuclear Non-proliferation Treaty', *Australian Journal of Politics and History* 50, no. 4 (2004): 526–43.

30 Scott D. Sagan, 'Why Do States Build Nuclear Weapons? Three Models in Search of a Bomb', *International Security* 21, no. 3 (Winter 1996/97): 63–4.

31 Graham T. Allison, *Essence of Decision: Explaining the Cuban Missile Crisis* (Boston: Little, Brown and Company, 1971), 67.

the nuclear domain, and exploring the feasibility of nuclear energy production.[32] This interest in the national development potentialities of nuclear technology was also reflected in Australia's diplomatic engagement on nuclear issues. Minister for External Affairs, H.V. Evatt, for example, in presenting Canberra's views with respect to the fledgling attempts of the newly formed United Nations Atomic Energy Commission to establish an international convention regulating nuclear technology in June 1946, noted Australia's interest not only the effective control of the military applications of such technology, but also the 'just and equitable' access to 'all information relevant to the peaceful use of atomic energy'.[33] The Chifley government injected funding into nuclear research to underpin these plans with a particular focus on investing in a strong capacity to drive this research in the university sector. The centrepiece of this strategy was the establishment of the Research School of Physical Sciences at the newly created Australian National University.[34] The Chifley government also created the first bureaucratic structures to coordinate defence and national development initiatives in the area of nuclear research, when it formed the Atomic Energy Policy Committee (later renamed the Industrial Atomic Energy Policy Committee) in 1949. The Committee continued under the Menzies government and was the forerunner of the Australian Atomic Energy Commission that was established under the Atomic Energy Act in 1953.[35]

While much of the 1953 Atomic Energy Act was couched in terms that could be interpreted as favouring military-related research and development, the nuclear weapons option was never really paramount in the government's thinking. Indeed, it became clear to Menzies and his senior Ministers 'that atomic energy was no longer a "defence" subject since all the research taking place was for industrial, academic and other peaceful purposes'.[36] This was reinforced in 1956 when Menzies directed that responsibility for nuclear energy matters be transferred from the Ministry of Supply to the Ministry of National Development. At the same time, however, while peaceful exploitation of the atom was the primary focus of the Menzies government, the energy security imperative for developing domestic nuclear power was not as compelling as it had been previously. By the mid-1950s, a better outlook was prevailing in terms of Australia's projected conventional fuel resources, which blunted the earlier

32 Reynolds, *Australia's Bid for the Atomic Bomb*, 48.

33 'Cablegram from Evatt to Department of External Affairs', in Reynolds and Lee, eds, *Australia and the Nuclear Non-Proliferation Treaty, 1945–1974*, 7.

34 Reynolds, *Australia's Bid for the Atomic Bomb*, 50–51.

35 For background, see Anna Eugenia Binnie, 'From Atomic Energy to Nuclear Science: A History of the Australian Atomic Energy Commission' (unpublished Doctor of Philosophy Thesis, Macquarie University, 28 February 2003), chapter 3.

36 Binnie, 'From Atomic Energy to Nuclear Science: A History of the Australian Atomic Energy Commission', 99.

sense of urgency surrounding uranium exploration and mining.[37] By the end of the 1950s, it had also become clear that major uranium markets in the United States and Britain were drying up due to over-supply. As a consequence, prices quickly dropped and the Australian uranium sector began to contract.[38]

The key international development that led to the deepening of Australia's engagement in the nuclear fuel cycle was the announcement in 1953 by the Eisenhower administration of the US 'Atoms for Peace' plan. The plan emphasized the great potential of nuclear technology to contribute to scientific advancement and the economic well-being of countries in the developing world. By harnessing nuclear energy for 'constructive, not destructive' purposes, the international community would be able to exploit the scientific benefits of the atom, which would be transferred to those countries that had not previously had access to nuclear technology.[39] Nuclear power was portrayed as the way of the future and would, in the words of senior US Atomic Energy Commission official Lewis Strauss, 'be too cheap to meter'.[40]

In Australia, research into an indigenous beryllium-oxide moderated high-temperature gas-cooled reactor design was an early focus of the AAEC during the 1950s and early 1960s, as it was seen as adaptable for smaller reactors more aligned with the needs of Australia's power system. In the 1960s, research then focused on fuel fabrication and zirconium technology in support of overseas, heavy water moderated natural uranium reactor designs, which also provided the option of value-adding to possible uranium exports. However, the AAEC never went so far as to operate even a prototype power reactor. Instead, the High Flux Australian Reactor (HIFAR, 10MW thermal) operated as a neutron source for material science purposes at Lucas Heights from 1960, complemented by the smaller MOATA (100kw thermal) reactor, also for use as a neutron source for scientific studies, from 1961.[41]

As soon as the Atoms for Peace initiative was promulgated, attention quickly turned to how the balance would be struck between the diffusion of nuclear technology and controls to ensure that it would not be used for military purposes.

---

37 Cawte, *Atomic Australia*, 97.

38 Geoffrey Blainey, *The Rush that Never Ended: A History of Australian Mining*, 3rd edition (Melbourne: Melbourne University Press, 1978), 338.

39 Richard Hewlett and Jack Holl, *Atoms for Peace and War, 1952–1961: Eisenhower and the Atomic Energy Commission* (Berkeley: University of California Press, 1989), chapter 8.

40 David Bodansky, *Nuclear Energy: Principles, Practices, and Prospects*, 2nd edition (New York: Springer-Verlag, 2004), 32.

41 Nuclear Energy Agency, *Trends in Nuclear Research Institutes* (Paris: OECD, 1996), 119; Committee of Review of the Australian Atomic Energy Commission, *Review of the Australian Atomic Energy Commission* (Canberra: Australian Government Publishing Service, 1986), 68.

Crafting the technical means to verify that nuclear technology being exported was not exploited to build weapons would be central to this achieving this balance. The result was the creation of the International Atomic Energy Agency (IAEA) in 1957 whose dual purpose was to promote the take up of nuclear energy worldwide, and to perform a watchdog role through a system of safeguards covering nuclear transfer arrangements.[42] An important part of the US-led initiative to institute the IAEA was Washington's close consultation with 'seven countries that had either developed raw material resources or maintained advanced atomic energy programs – namely, the UK, France, Canada, Australia, Belgium, South Africa, and Portugal'.[43] Australia's involvement in drafting the statute of the IAEA as part of the consultative group derived specifically from its 'recognised potential to produce and supply substantial quantities of source material, including uranium'.[44] This was a natural continuation of its close engagement in international nuclear diplomacy, which began with involvement in the UN Atomic Energy Commission from 1946 until its dissolution in 1952. Following the IAEA's inception in 1957, Australia was appointed to the Agency's Board of Governors on the basis of being identified as the most advanced state in the Southeast Asian and Pacific regions in terms of its nuclear scientific and technical expertise.

Despite this, however, successive governments from Menzies to McMahon were apprehensive about the impact of nuclear safeguards on Australia's interests. Primarily this was due to concern over the potential for commercial and military espionage on the part of international inspectors visiting Australian nuclear sites. Uncertainty over the scope and authority that would be granted to the IAEA and its inspectors under the new NPT regime worried Australian policy makers and raised questions about national sovereignty issues, including, for example, whether Australia would have the right to exclude inspectors who were nationals of Communist bloc countries.[45] Yet, even in this context the weapons dimension was less of a concern than commercial considerations in relation to securing unfettered Australian access to global uranium markets and the related capacity to undertake indigenous research and development activities, including on enrichment technologies.

A Department of Defence 'note for file' on President Eisenhower's proposal for the establishment of the IAEA of December 1953, for example, noted that Australia could be a potential contributor to the agency via the provision

42 See Lawrence Scheinman, *The International Atomic Energy Agency and World Nuclear Order* (Washington, DC: Resources for the Future, 1987).

43 Hewlett and Holl, *Atoms for Peace and War*, 229.

44 John Stoessinger, 'Atoms for Peace: The International Atomic Energy Agency', in Commission to Study the Organization of Peace, ed., *Organizing Peace in the Nuclear Age* (New York: New York University Press, 1959), 119.

45 On this point, see Bellany, *Australia in the Nuclear Age*, 106–8.

of uranium and as a potential recipient of international nuclear expertise with respect to the 'peaceful uses' of nuclear technology.[46] While Australian uranium was already exported under bilateral agreements with the United States and United Kingdom, the note expressed some doubt as to whether Australia would be able to engage in any further 'downstream' processing, such as enrichment, if it were to accede to such an international body. The note also suggested however that Australia 'could well claim' assistance from such an international body on the basis that it was 'power starved in the sense used by President Eisenhower in that we lack alternative sources', 'we have established and are developing a nucleus of scientific facilities ... personnel and equipment' and 'we could physically construct, operate and benefit from atomic power-houses'.[47]

Concerns about international inspections were again raised in the 1960s with the government considering supporting plans by Rio Tinto for cooperation between Australian uranium producers and the uranium fuel fabricating industry in Japan.[48] In October 1961, Minister for National Development W.H. Spooner informed Prime Minister Menzies that Rio Tinto believed that as most countries except for the United States and Soviet Union 'will be able to support only one fuel element fabrication plant each', the interests of uranium producers lay in 'a tie-up with plants for uranium metal manufacture, enrichment and nuclear fuel fabrication'.[49] The company, Spooner reported, also foresaw good potential to lock in a long-term consumer by noting that Japanese demand for uranium oxide was estimated to 'approach an order of 1,000 tons per annum between about 1968 and 1980' and that Japan was likely to develop an industrial enrichment capacity so as to not be 'heavily dependent' on foreign suppliers and facilities.[50] Yet, Spooner was also quick to note that the development of such a relationship could place Australia in a dilemma with respect to potential proliferation:

> Any dilemma which might face us in this matter stems from the fact that on the one hand countries like Japan and India could provide substantial markets for Australian uranium oxide. On the other hand, production of plutonium from natural uranium fuelled reactors and the development of facilities for enrichment of uranium, both provide the source materials for the production of nuclear weapons.[51]

---

46 'Note for File by Department of Defence: President Eisenhower's Proposal for an International Atomic Energy Agency', December 1953, in Reynolds and Lee, eds, *Australia and the Nuclear Non-Proliferation Treaty, 1945–1974*, 15.

47 Ibid., 16–17.

48 Letter from the Secretary, Prime Minister's Department, entitled 'Rio Tinto Mining Company', 1961, NAA A1209, 3052581.

49 Letter from Minister for National Development W.H. Spooner to Prime Minister Robert Gordon Menzies, 23 October 1961, NAA A1209, 3052581.

50 Ibid.

51 Ibid.

National development and potential defence requirements also impinged on government consideration of the prospect for Australian uranium supply to Japan throughout the remainder of the decade. In January 1967 Victorian Premier Henry Bolte, as his state embarked on 'a thorough investigation of nuclear power', sought assurances from the Holt government that 'any export agreements that may be entered into … are subject to the overriding condition that Australian requirements are to be satisfied first'.[52] The Department of Trade and Industry advised that the Prime Minister 'would be able to reassure Sir Henry Bolte' that the government's policy of prohibiting export of uranium except with the approval of the AAEC was 'to ensure that reasonable stocks of uranium are retained in Australia for any foreseeable defence or nuclear power requirements'.[53]

Reluctance to commit to strengthening international safeguards reached a high point in the late 1960s when it looked as though the planned expansion of nuclear power production in Europe, Japan and the United States would trigger a significant demand for low-cost Australian uranium in the longer term.[54] It would not be until 1974 that Australia signed a 'model' safeguards agreement with the IAEA, fulfilling its basic obligations as a non-nuclear weapons state under the NPT.[55]

The intersection between Australian governments regarding nuclear energy as a potential military instrument and as a tool of national development culminated in the late 1960s with Prime Minister John Gorton's announcement in 1969 that 'the Commonwealth Government would take Australia into the atomic age by beginning the construction of an atomic plant at Jervis Bay [in New South Wales] to generate electricity'.[56] The nuclear reactor would provide Australia with a nuclear infrastructure well beyond the limited capacity of the experimental reactor at Lucas Heights. Senator Robert Cotton, representing the Minister for National Development, explained in the Senate that

---

52 Letter from Premier Henry Bolte to Prime Minister Harold Holt, 20 January 1967, NAA A1209, 3052581.

53 Letter from the Department of Trade and Industry on the export of uranium to Japan, 7 February 1967, NAA A1209, 3052581.

54 Ciaran O'Faircheallaigh, 'Uranium in Australia, 1970–1987: An Overview', in Ciaran O'Faircheallaigh, Deborah Wade-Marshall and Ann Webb, eds, *Uranium in Australia, 1970–1987: An Annotated Bibliography* (Canberra: Australian National University, North Australia Research Unit, 1989), 12–13.

55 Australian Department of Foreign Affairs, 'Agreement between Australia and the International Atomic Energy Agency for the Application of Safeguards in Connection with the Treaty on the Non-Proliferation of Nuclear Weapons: Entry into Force, 10 July 1974', *Treaty Series 1974*, 16 (Canberra: Australian Government Publishing Service, 1974).

56 Binnie, 'From Atomic Energy to Nuclear Science: A History of the Australian Atomic Energy Commission', 208.

> The Government wishes Australia to obtain experience in ordering, commissioning and operating nuclear power stations and also wishes to assist in establishing a nuclear industry so that Australia will be in a favourable position to adopt fast breeder reactors when they become commercially available.[57]

Such ambitions reflected Gorton's own nationalist instincts for genuine self-reliance, which complemented the views of the AAEC and its influential chairman, Sir Philip Baxter.[58] As Alice Cawte has outlined, the Jervis Bay initiative was driven by a mixture of aspirations, including maximizing the national economic benefits of nuclear energy, adding value to uranium exports through enrichment and also potentially achieving a threshold nuclear weapons option.[59]

Between 1969 and 1972, when the project was shelved indefinitely following Gorton's political demise, significant preparations were put in place by the AAEC to have the reactor operational by 1975. However, the pressurized water design that was ultimately chosen for the reactor was the least useful of those under consideration, should the reactor ever have been used for plutonium production.[60] In many respects, the ill-fated Jervis Bay project captured the essence of Australia's postwar nuclear policy. The intersection of military and economic considerations were fundamental to the way in which successive Australian governments from 1945 to 1972 perceived opportunities surrounding nuclear energy. But Australia never had an active nuclear weapons program, and any movement towards acquiring weapons directly from the US or UK or via national indigenous production were always tentative. Even the Jervis Bay project was seen by pro-acquisition constituencies as potentially providing a future threshold capability rather than a nuclear weapons arsenal for Australia, and rose and fell on economic considerations. By the early 1970s, any earlier imperative that existed for the domestic production of nuclear energy floundered on the realization that Australia enjoyed massive fossil fuel advantages that offset any market advantages nuclear energy may have in generating electricity – Cabinet halted the Jervis Bay project in 1971 following advice that the cost of nuclear electricity would be 30% higher than from a conventional power plant.[61] It would not be until the Howard government – some three and a half decades later – that Australia would experience another serious discussion about the merits and disadvantages of developing a domestic nuclear power industry.

---

57 See remarks by Rt. Hon. Robert Cotton, *C.P.D.* (Senate), *Atomic Energy*, 1 October 1970, 1045.

58 On Baxter's role in shaping Australia's postwar nuclear policy, see Roy MacLeod, 'The Atom Comes to Australia: Reflections on the Australian Nuclear Program, 1953 to 1993', *History and Technology* 11, no. 2 (1994): 304–9.

59 Cawte, *Atomic Australia*, 127–30.

60 Hymans, 'Isotopes and Identity: Australia and the Nuclear Weapons Option, 1949–1999', 10–11.

61 Cabinet Decision No. 195, 8 June 1971, NAA A5908, 121.

Australia and the Non-Proliferation Treaty

As the most widely observed arms control treaty since the Second World War, the NPT provides the legal and political framework of the international nuclear non-proliferation regime.[62] The product of a joint American-Soviet draft treaty submitted to the United Nations General Assembly (UNGA) on 24 August 1967, the NPT was opened for signature on 1 July 1968 with minor amendments and entered into force on 5 March 1970. Defining a Nuclear Weapons State (NWS) as 'one which has manufactured and exploded a nuclear explosive device prior to January 1, 1967', the Treaty mandates a distinction between those states deemed to legitimately possess nuclear weapons and those Non-Nuclear Weapons States (NNWS) who have forsworn the acquisition of nuclear weapons for as long as they remain party to the NPT. The core of the Treaty's non-proliferation provisions are contained in Articles I and II which proscribe the transfer of nuclear weapons technology by NWS to NNWS and the receipt of such technology and the manufacture of nuclear weapons by the latter. In committing themselves to the non-acquisition of nuclear weapons, NNWS are required under Article III to conclude safeguards agreements with the International Atomic Energy Agency (IAEA) to verify the non-diversion 'of nuclear energy from peaceful uses to nuclear weapons or other nuclear explosive devices'.

The quid pro quo contained in the NPT is a commitment by NWS that they will facilitate 'the fullest possible exchange of equipment, material, and scientific and technological information for the peaceful uses of nuclear energy' (Article IV), and work towards 'a treaty on general and complete disarmament under strict and effective international control' (Article VI). While critics of the NPT have argued that its inherently discriminatory provisions reinforce the already privileged status of the five declared nuclear powers, defenders of the Treaty have countered that the division between NWS and NNWS fosters a sense of confidence (and thus stability) among states that the number of countries openly possessing nuclear weapons will remain frozen at 1967 levels.[63]

To compensate for the absence of a provision in the Treaty establishing an executive arm to oversee its implementation, member states have conducted NPT Conferences at five-yearly intervals. Review Conferences comprise several plenary sessions at which official representatives issue policy statements, a number of Main Committees in which the implementation of the Treaty's

62 The text of the Treaty is reproduced in United Nations, *Status of Multilateral Arms Regulation and Disarmament Agreements*, 3rd edition (New York: United Nations Department for Disarmament Affairs, 1988), 71–6.

63 For contrasting perspectives, see Ashok Kapur, 'The Future of the NPT: A View from the Indian Subcontinent', in D. Dewitt, ed., *Nuclear Non-Proliferation and Global Security* (New York: St. Martin's Press, 1987), 201–15; Joseph S. Nye, 'NPT: The Logic of Inequality', *Foreign Policy* 59 (Summer 1985): 123–31.

various Articles is reviewed, and a Drafting Committee to formulate compromise wording on a consensus Final Declaration. While substantive issues have often limited the scope for agreement, NPT Conferences have provided a forum for member states to evaluate the efficacy of IAEA safeguards, propose additional measures to strengthen the operation of the Treaty and monitor the nuclear policies of non-NPT states.[64]

Since the Whitlam government's ratification of the NPT in January 1973, successive Coalition and Labor governments have reaffirmed Australia's strong commitment to the Treaty. However, this followed a period of equivocation within official circles between 1968 and 1972 over whether Australia should accede to the NPT. Although the interlude between the opening of the NPT for signature in 1968, Australia's signature of the Treaty in February 1970 and ratification of the Treaty in 1973 was not especially protracted in relative terms,[65] it did reflect certain doubts within the Gorton and McMahon Coalition governments regarding the merits of Australia becoming an NNWS under the NPT. Before outlining Australia's pattern of support for the NPT after 1973, it is worth pausing briefly to consider why Australia was initially reluctant to commit itself to the Treaty.

The question of whether Australia should sign the NPT provoked modest, though at times animated, debate among official and independent observers between 1967 and 1970. In the period immediately preceding, and in the two years after, the opening of the NPT for signature, the attitude of the Australian government towards the Treaty was characterized by guarded scepticism. As Richardson noted at the time, although official declarations by the Gorton government tended to highlight the 'reasons why [the NPT] may prove ineffective', they accorded scant attention to 'the Treaty's potential value as a non-proliferation measure'.[66] With respect to the former, a major issue for Canberra was whether the conclusion of the NPT would adversely affect US security assurances to Australia. For example, Australia's ambassador to the US, Keith Waller, pressed officials in the State Department in November 1967 on the US response to a 'hypothetical' conventional attack on Australia by a 'non-nuclear power that had signed the NPT and did not receive assistance in the attack from a nuclear power'.[67] The State Department response was that 'if the

64 J. Goldblat, *Twenty Years of the Non-Proliferation Treaty: Implementation and Prospects* (Oslo: International Peace Research Institute, 1990), 50–53.

65 Indonesia, for example, signed the NPT in 1970, but did not deposit its instrument of ratification until 1979.

66 J.L. Richardson, 'Australia and the Non-Proliferation Treaty', *Canberra Papers on Strategy and Defence*, no. 3 (Canberra: Australian National University Press, 1968), 10.

67 'Cablegram from Waller to Department of External Affairs: Non-Proliferation – Security Assurances', 29 November 1967, in Reynolds and Lee, eds, *Australia and the Nuclear Non-Proliferation Treaty, 1945–1974*, 157.

US used conventional weapons on a massive scale to stop aggression on the peninsulas of Asia (Korea and Vietnam), there should be no doubt at all that it would do so in the case of Australia' and that ultimately 'if supreme national interests were endangered' Australia could always withdraw from the NPT. 'I am forced to conclude', the ambassador noted, that there thus 'may be some reduction in the value of American nuclear deterrent as a factor in our security'.[68]

Shortly before the conclusion of the NPT in July 1968, the Gorton government articulated Australia's principal concerns with the draft treaty then being debated at the UNGA. In March 1968 External Affairs Minister Paul Hasluck stated that while the government desired 'an effective and equitable treaty on the non-proliferation of nuclear weapons', given that 'the treaty may have considerable effects on the future development of Australia ... a final and detailed position cannot be stated until there has been much further examination'.[69] Hasluck noted that particular attention would be paid to 'the impact on Australia of an effective and pervasive inspection system, and the degree to which Australia would be bound by future international decisions under the draft treaty regarding inspection or other controls'.[70] Minister for National Development David Fairbairn, in a letter to Prime Minister Gorton, perhaps best expressed such concerns when he asserted that, 'There is little doubt that the IAEA Safeguards System, if applied universally will be very costly to the Commonwealth and is likely to have an adverse effect on the introduction of nuclear power and development of associated industries in this country'.[71]

The concerns of the government were elaborated in more detail two months later when Australia's Ambassador to the United Nations addressed the General Assembly. Punctuated by misgivings about the draft treaty's potential impact upon state sovereignty, Shaw's statement set out the Coalition's non-committal policy of neither rejecting the NPT outright nor legally committing Australia to abide by its provisions.[72] The statement emphasized the government's apprehension concerning the possible future obstacles any treaty could present to Australia's utilization of nuclear explosives for large-scale engineering projects and the potential scope for espionage resulting from compliance with IAEA safeguards on indigenous nuclear activities.[73] Significantly, in acknowledging the right of withdrawal clause contained in the draft treaty, the statement hinted that Australia would not foreclose the option of acquiring nuclear weapons even

68 Ibid., 158.

69 *C.P.D.* (Representatives), 26 March 1968, 456.

70 Ibid.

71 'Letter from Fairbairn to Gorton', 16 February 1968, in Reynolds and Lee, eds, *Australia and the Nuclear Non-Proliferation Treaty, 1945–1974*, 173.

72 'Draft Treaty on the Non-Proliferation of Nuclear Weapons', *Current Notes on International Affairs* (Statements) 39, no. 5 (May 1968): 206–10.

73 Ibid., 208–10.

after it had acceded to any operative non-proliferation treaty.[74] Indeed, Charles Mott, Acting Director of the Political Affairs Section of the UN Branch of the Department of External Affairs, noted in this respect that it 'would … be open to us, once we had invoked Article X, to seek weapons from others' and that 'the Treaty would not prevent our engaging in the development of missiles which would have potential use for the delivery of nuclear warheads'.[75]

Although the decision by the Gorton government to sign the NPT in February 1970 did not legally bind Australia to the terms of the Treaty,[76] it did provide supporters of Australian accession, such as the Labor Opposition, with a foundation on which to base their case for immediate Australian ratification. Australia's eventual ratification of the Treaty in January 1973 was a direct consequence of Labor's 1972 pre-election pledge to legally validate Australia's status as an NNWS under the NPT.[77] Linking ratification with his government's 'strong and unequivocal support for efforts to check the spread of nuclear weapons',[78] Whitlam portrayed formal backing for the Treaty as a corollary of Australia's aim of 'doing all we can as a medium-sized power to help all nations … make the most of the new [international] opportunities now presenting themselves'.[79]

During its brief tenure, the Whitlam government's support for the NPT served to bolster Australia's diplomatic credibility in its more high-profile forays into international nuclear affairs.[80] Throughout Labor's period in office, Australia's policy concerning the NPT was dual-track in nature. The first element was an acknowledgment of a correlation between the nuclear policies of the NWS and the future prospects for nuclear non-proliferation. As Australia's representative to the first NPT Review Conference argued:

---

74 Ibid., 208.

75 'Minute from Mott to Pritchett and Thomson: Draft Treaty on the Non-Proliferation of Nuclear Weapons', 6 February 1968, in Reynolds and Lee, eds, *Australia and the Nuclear Non-Proliferation Treaty, 1945–1974*, 165.

76 In announcing the government's decision to sign the NPT, Gorton pointedly remarked that it was 'not to be taken in any way as a decision to ratify the Treaty, and of course, the Treaty is not binding on us until it is ratified'. 'Nuclear Non-Proliferation Treaty', *Current Notes on International Affairs* (Statements) 41, no. 2 (February 1970): 70.

77 Speech delivered by then leader of the Australian Labor Party, Edward G. Whitlam, MP, 'Australian Labor Party Policy' (Blacktown Civic Centre, 13 November 1972), 32.

78 'Ratification of the Nuclear Non-Proliferation Treaty', *Australian Foreign Affairs Record* (Statements) 44, no. 2 (February 1973): 129.

79 *C.P.D.* (Representatives), 24 May 1973, 2644.

80 The most conspicuous of which was Australia's joint prosecution, with New Zealand, of a case against French atmospheric nuclear testing in the South Pacific at the International Court of Justice between 1973 and 1974.

> We are convinced that, if the nuclear-weapon powers do not themselves show restraint, non-nuclear-weapon powers will also be disinclined to show restraint. Unless the nuclear powers demonstrate their own good faith by forswearing progressively their own nuclear arsenals, it will be that much harder to persuade non-nuclear powers to forswear altogether their nuclear weapons options. Horizontal and vertical proliferation are inextricably connected. All parties must show an equal devotion to the non-proliferation regime.[81]

Stressing the causal relationship between vertical and horizontal proliferation provided a broader rationale for the Whitlam government's claim regarding the necessity of a Comprehensive Test Ban treaty to ensure the longer-term coherence and effectiveness of the NPT as a non-proliferation instrument.[82]

The second policy track adopted by the Whitlam government was an endeavour to make the NPT more universal in its application to nuclear and non-nuclear weapon states alike. This was reflected in a number of statements that underlined the importance of larger powers in the Asia-Pacific (particularly China and Japan) and Europe (such as France and West Germany) acceding to the Treaty as a way of encouraging smaller states to follow suit.[83] The Whitlam government's most lasting impact on Australia's non-proliferation diplomacy, however, was indirect, through the creation of the Ranger Inquiry that would report to the Fraser government that succeeded it.

Deferring Decisions: Whitlam and the Ranger Inquiry

Between 1970 and 1972, four major new deposits of uranium were discovered at Nabarlek, Ranger, Koongarra and Jabiluka in the Northern Territory, estimated to be some of the largest in the world.[84] The AAEC resumed its major role in uranium exploration, and by the time Whitlam came to power at the end of 1972, it reported that Australian uranium reserves comprised roughly 20% (188,000 tonnes) of the total estimated world resources. This was a period in which global demand for uranium was set to soar, on the back of increasing numbers of power production reactors worldwide, marking the end of cheap world uranium prices. As the business case for nuclear power production in Australia faltered in the early 1970s, the national development potential of

---

81 'Swift Action Needed to Curb Proliferation: Text of a Statement Made on 8 May 1975 by Mr R.W. Furlonger, Leader of the Australian Delegation to the Review Conference of the Parties to the Treaty on the Non-Proliferation of Nuclear Weapons', *Australian Foreign Affairs Record* (Statements) 46, no. 5 (May 1975): 280–81.

82 See Whitlam's comments in *C.P.D.* (Representatives), 26 August 1975, 483.

83 See for instance 'Swift Action Needed to Curb Proliferation …', 279; Whitlam's comments in *C.P.D.* (Representatives), 11 February 1975, 63.

84 Cawte, *Atomic Australia*, 136–7.

nuclear industry had narrowed to Australia's uranium reserves – but these now seemed to promise great rewards if they could fuel the growing nuclear power industries abroad.[85]

Although the Whitlam government ratified the NPT, it had also won office with a pro-uranium mining and development stance in its platform. Whitlam's Minister for Minerals and Energy, Reginald 'Rex' Connor, sought to capitalize on the rise in demand through increasing the value of Australia's uranium exports by enriching it in Australia first.[86] Research into uranium conversion and centrifuges for enrichment became important AAEC priorities during the 1970s.[87] Connor tightened procedures for exports by taking the decision of approval of new export contracts away from the AAEC and giving it to his own Department, which gave him the power to refuse approval of any new export contracts of uranium. Given a prediction that world uranium prices would continue to rise to around $100 per kilo in the early 1980s, at which time Australia would be able to sell enriched uranium on the market for a much higher price than would be possible for yellowcake, Connor wanted to freeze Australian sales until such time. In the interim, he set about developing a framework for government-industry joint cooperation in mining of the Northern Territory Ranger deposit, and held talks with the AAEC and their Japanese counterpart over the potential development of an enrichment plant in Australia.[88]

The desire to reap the projected economic windfall of this 'uranium boom' was, however, complicated for the Whitlam government by its pre-election commitment to advance the land rights of Australia's Aboriginal peoples.[89] These goals – maximizing the economic potential of uranium and advancing Aboriginal land rights – thus intersected with some contradiction. Therefore, Australia's ratification of the NPT as an NNWS, the promise of a potential uranium 'boom', and Aboriginal land rights converged to generate domestic political debate on whether or not Australia should export uranium, and if so, on what basis.

Following its election, the Whitlam government delayed taking a decision on the future of Australian uranium exports pending the negotiation of terms with companies based in Australia to ensure large-scale Commonwealth involvement

---

85 Broinowski, *Fact or Fission?*, 121–2.

86 The Whitlam government's desire for a uranium enrichment capability is examined in detail in Wayne Reynolds, 'Australia's Quest to Enrich Uranium and the Whitlam Government's Loans Affair', *Australian Journal of Politics and History* 54, no. 4 (2008): 562–78.

87 Nuclear Energy Agency, *Trends in Nuclear Research Institutes*, 119.

88 Broinowski, *Fact or Fission?*, 121–2.

89 Jason O'Brien, 'Canberra Yellowcake: The Politics of Uranium and How Aboriginal Land Rights Failed the Mirrar People', *Journal of Northern Territory History* 14 (2003): 80.

in mining projects. In October 1974 agreement was reached between the government and Peko Mines Limited and the Electrolytic Zinc Company over the development of the Ranger uranium deposit in the Northern Territory, with the Commonwealth acquiring 50% of the Ranger project at a cost of financing 72.5% of its development.[90] Yet, the Whitlam government subsequently made it clear to the joint venture partners that it would not approve the Ranger venture until it had considered the findings of the Ranger Uranium Environmental Inquiry which it created on 16 July 1975.[91]

The Ranger Inquiry (or, as it was also referred to after its Presiding Commissioner Justice Russell Fox, the Fox Commission)[92] began hearing evidence on 9 September 1975.[93] The Inquiry was constituted under the provisions of the *Environment Protection (Impact of Proposals) Act*, and the terms of reference of the Inquiry extended to an investigation into all environmental aspects of the joint Commonwealth-private sector Ranger proposal. The first of its type in Australia, the Inquiry was established under a statutory instrument of government and therefore not subject to ministerial direction. Consequently, the Inquiry's Commissioners retained considerable discretion in determining the time taken with proceedings, the nature of evidence collected and the very scope of the Inquiry itself. While the Ranger Inquiry met formally for just over four months, it received evidence from 303 witnesses, received 419 separate exhibits, and the Report took some 11 months to complete.[94] Despite the ostensibly strict terms of reference provided by the *Act* under which it was constituted, the Inquiry came to be regarded by many as an authoritative (and impartial) investigation into the implications of Australian involvement in the international nuclear fuel cycle as a supplier of uranium.[95] Indeed, declarations from the newly elected Coalition government that announcement of its uranium export policy would be left until after the presentation of the Inquiry's

---

90 See 'Memorandum of Understanding: Parties to this Memorandum – The Commonwealth of Australia - (Australia), Peko Mines Limited - (Peko), Electrolytic Zinc Company of Australasia - (EZ)', tabled in *C.P.D.* (Senate), 26 October 1977, 1760–63.

91 Edward G. Whitlam, *The Whitlam Government: 1972–1975* (Ringwood: Penguin Books, 1985), 544.

92 The other two Commissioners were Professor Charles Kerr and Graeme Kelleher.

93 'NT Uranium Project "New Basic Industry"', *The Canberra Times*, 10 September 1975.

94 The procedures of the Inquiry were semi-judicial in nature: all evidence was required to be given in public on oath or affirmation and every witness retained the right to cross-examine every other witness.

95 It should be noted that on 11 November 1976 the Fraser Government granted permission for the early delivery of Australian uranium under existing export contracts from the Mary Kathleen mine in Queensland. See 'Mary Kathleen Uranium: Export Permits Allowed', *The Canberra Times*, 12 November 1976.

findings[96] imbued the Inquiry with an aura of authority which it would not have otherwise possessed.

The First Report of the Ranger Inquiry was presented to the Fraser government on 28 October 1976.[97] In a wide-ranging, equivocal and at times contradictory evaluation of contemporary international nuclear developments, the Report's analysis was divided into two parts. The first part dealt with the scientific attributes of the nuclear fuel cycle, the role of uranium in the development of nuclear power and the economic vagaries of international uranium supply and demand. The second part of the Report looked more explicitly at the potential risks arising from the use of nuclear power. Apart from the environmental hazards resulting from nuclear reactor accidents and the disposal of high level nuclear waste, the Report identified the risks of possible theft of weapons-grade nuclear material by non-state terrorist organizations from national stockpiles.[98]

Yet by far the greatest attention of the Report was devoted to the issue of nuclear proliferation among states. The Report highlighted what it saw as the 'inherent conflict of aims' embodied in the IAEA Statute and the NPT regarding, on the one hand, the promotion of nuclear energy in its 'peaceful' forms and, on the other hand, the objective of preventing the further spread of nuclear weapons capability to individual countries.[99] The Report was especially critical about the question whether existing international safeguards provided an adequate barrier to states acquiring nuclear weapons. Noting boldly that safeguards 'do not hinder a country wishing to develop nuclear weapons', the Report proceeded to summarize what the Commissioners saw as the main shortcomings of existing international safeguards:

> the failure of many states to become parties to the NPT; the inability of safeguards to prevent the transfer of nuclear technology from nuclear power production to the acquisition of nuclear weapons competence; the fact that many nuclear facilities are covered by no safeguards; the existence of a number of loopholes in safeguards agreements regarding their application to peaceful nuclear explosions, to materials intended for non-explosive military uses, and to the retransfer of materials to a third state; the absence, in practice, of safeguards for source materials; the practical problems of maintaining effective checks on nuclear inventories; the ease with which states can withdraw from the NPT and from most non-NPT safeguards agreements; deficiencies in accounting and

---

96 See the comments by Deputy Prime Minister Doug Anthony in *C.P.D.* (Representatives), 4 March 1976, 526.

97 Russell W. Fox, *Ranger Uranium Environmental Inquiry: First Report* (Canberra: Australian Government Publishing Service, 1976).

98 Ibid., 151.

99 Ibid., 126.

> warning procedures; and the absence of reliable sanctions to deter diversion of safeguarded material.[100]

These flaws were, according to the Report, sufficiently serious to warrant the conclusion that existing safeguards provided merely 'an illusion of protection' against nuclear weapons proliferation.[101]

The Report concluded that in the event of Australia transferring uranium, the government should unilaterally apply bilateral export pre-conditions in addition to those already incorporated in NPT-type (INFCIRC/153) safeguards agreements. The most important would be a requirement for *prior* existing full-scope safeguards over all nuclear material within the prospective recipient state. In order to strengthen the effect of this export pre-condition, the Report outlined three additional recommendations. First, that any agreement entered into by the Australian government to supply uranium include a commitment by the recipient party that no nuclear material be employed for military purposes irrespective of any alteration to their bilateral safeguards agreement with the IAEA. Second, that any transfer agreement incorporate a clause preventing the recipient country from retransferring Australian nuclear material to a third party without first gaining the Australian government's approval; this, it was presumed, would prevent the possibility of the recipient state retransferring Australian material to another state under less stringent safeguards conditions. Finally, the Report recommended that, in order to circumvent the possibility of Australian nuclear material no longer being subject to safeguards in the event of a recipient country withdrawing from the NPT or from its bilateral agreement with the IAEA, the Australian government should only supply material to those states whose 'entire nuclear industry is subject to back-up safeguards that cannot be terminated by unilateral withdrawal'.[102]

Although the Ranger Inquiry had been established under the *Environment Protection (Impact of Proposals) Act*, the commission interpreted its brief in a most wide-ranging manner. In essence, it approached the question of uranium exports as one of Australia's nuclear policy as a whole, focusing on the complex links between the normative and economic domains of nuclear policy. How to craft a workable nuclear policy from these considerations in a shifting international context became a central question for the new Fraser government.

Conclusion

Australia's engagement with nuclear issues from 1945 to 1975 was characterized by the development of both complementarities and tensions between

100 Ibid., 147.

101 Ibid.

102 See ibid., 147–9.

Canberra's interests in each domain of nuclear policy. In the strategic domain, Australian political leaders and policy elites actively considered the acquisition of nuclear weapons. Crucially, consideration of this question was framed by evolving perceptions of not only the global and regional security environment but also by competing understandings of the 'ordering' role of nuclear weapons in international affairs, and the relevance of nuclear deterrence and arms control to Australian security. As demonstrated in this chapter, the 'Menzian' vision of these issues, that viewed nuclear weapons as a potential stabilizing factor in international affairs so long as they remained wielded by 'responsible' great powers, came to dominate Australian views (with the exception of the brief tenure of Prime Minister John Gorton) of the strategic domain during this period.

This position clearly impacted on Australian consideration and assessment of emerging arms control initiatives from the early 1960s onward and contributed to Canberra's decision to sign and ratify the NPT as an NNWS. However, as this chapter has demonstrated, Australia's equivocation between 1968 and 1972 over accession to the NPT as an NNWS was driven by concerns that such a commitment would constrain the country's freedom of action with respect to the strategic and economic applications of nuclear technology. Ultimately, Australian leaders and policy makers judged that the structure and logic of the NPT – limiting nuclear arsenals to the five existing NWS in 1968 and establishing mechanisms to encourage proliferation restraint amongst NNWS members – best served Australia's security interests.

The impact of NPT membership on Australia's ability to harness nuclear technology for national development goals was however more ambiguous and contributed to tensions between Canberra's interests in the strategic, normative and economic domains. Article IV enshrined NNWS members' 'inalienable right' to access and develop nuclear technology for 'peaceful purposes', establishing a complementarity between Australia's strategic interests in a potential weapons option and national development goals by providing the conditions for the full development of nuclear infrastructure and fuel cycle technology through the AAEC's mandate. However, Australia's commitment to do to so under IAEA safeguards and inspections (Article III of the NPT) prompted some concern that opening Australian nuclear facilities to such inspections could risk commercial and military espionage. Finally, while NPT membership also enabled the export of Australian uranium (via the 'peaceful use' provisions of the NPT), such a clear economic interest potentially conflicted with Canberra's strategic interest to not facilitate further nuclear proliferation. The question of how Australia could effectively navigate between such complementarities and tensions in its approach to each of the domains of nuclear policy was ultimately left for the Fraser government to answer.

Chapter 3
Bedding Down the NPT: The Fraser Government (1975–83)

The government of Malcolm Fraser came to power when Cold War confrontation was becoming increasingly tense again. Soviet military modernization in the 1970s, combined with the erosion of the US nuclear advantage, caused significant concern among many US allies, and Malcolm Fraser was no exception. The era of détente ended with the Soviet invasion of Afghanistan in 1979, and several years of high tension between Western countries and the Soviet Union followed that did not end before the Labor Party returned to office in Australia.

When Fraser became Prime Minister in 1975, the future of the NPT was far from assured. Whereas Australia had ratified the NPT under the Whitlam government in 1972, many other countries had not yet done so. Germany and Italy ratified the treaty in 1975, Japan in 1976, Switzerland in 1977, and, importantly for Australia, Indonesia did not ratify until 1979. Between the end of 1975 and the end of 1982, roughly Fraser's time as Prime Minister, NPT membership rose from 95 to 118 countries. Many regionally important or nuclear capable countries still had not acceded to the treaty in 1983, including France, China, India, Pakistan, Israel, South Africa, Algeria, Saudi Arabia, Brazil and Argentina.

Even among members to the treaty, negotiation of nuclear safeguards arrangements was a major factor behind the often-delayed ratification, and the development and implementation of safeguards for many nuclear technologies only had begun in 1975. European countries with major nuclear industries, as well as Japan, shared Australian concerns that the IAEA safeguards regime would discriminate against non-nuclear weapons states. Members of EURATOM had often made signature conditional on the operation of that treaty, which established free trade in nuclear materials among member states and a multilateral safeguards regime administered by EURATOM inspectors. A cornerstone of the European Economic Community system, EURATOM was of practical and symbolic importance to its members.[1] In general, the international politics of the nuclear fuel cycle within which Australia operated and the economic potential of Australia's uranium remained in flux, and would also have significant implications for the attempt to build a centrifuge-based enrichment program in Australia.

1 A fact that Australia would, in the end, have to accommodate in the bilateral safeguards provisions on re-transfer of Australian uranium.

Overall, the main task for the Fraser government in nuclear policy was thus to bed down the NPT – both in Australian policy and internationally – in regards to all three of its dimensions: nuclear weapons state obligations, safeguards and civilian use. Governing in challenging international circumstances, it developed policies that later governments would amend, but seldom overturn.

Australia's Security and the Global Nuclear Balance

When the Whitlam government had assumed office in 1972, there had been good reasons to hope that a stable global nuclear order would take shape in the 1970s. Washington's and Moscow's embrace of détente, the Anti-Ballistic Missile Treaty and the Strategic Arms Limitation Treaty (SALT) of 1972 suggested a moderation of the nuclear arms race. The NPT had been signed (if not always ratified) by many countries with a nuclear industry. The overthrow of Sukarno in Indonesia in 1965 and the emerging rapprochement between the United States and China in 1971–72 were also viewed as favourable developments in Canberra.[2]

Only a few years later, however, global security developments were cause for far greater concern. Globally, détente and super power arms control were stalling. The Soviet Union was modernizing its nuclear ICBM force, while MIRV technology and increased precision seemed to make a first strike more feasible than it had been in the past. The United States appeared to be in relative decline, and the value of US security guarantees eroded – a perception that could be traced to a range of factors including the 1968 Guam doctrine; reductions in US conventional forces; the fissures between the United States and its allies that had become apparent during the 1973 Middle East war; the Watergate scandal; Congressional aversion to international commitments, as evident in the War Powers resolution and the refusal of aid to South Vietnam in 1975; and the Schlesinger doctrine of limited nuclear warfighting options. At the same time, the Soviet Union demonstrated increasing international reach, including deployments of naval forces in the Indian Ocean; airlifts to its Arab allies in 1973; and deployments of Cuban proxies to Angola in 1975 and Ethiopia in 1977.[3]

In this context, the return to power of the Liberal-Country Party Coalition in November 1975 augured significant changes in Australian foreign policy.

2 See for example J.L.S. Girling, 'Australia and Southeast Asia in the Global Balance: A Critique of the "Fraser Doctrine"', *Australian Outlook* 31, no. 1 (1977): 3–6; Robert Catley, 'Australia and the Great Powers', *Australian Outlook* 37, no. 3 (1983): 144–5.

3 For Australian perspectives on these developments, see the essays in Robert O' Neill, ed., *The Strategic Nuclear Balance: an Australian Perspective* (Canberra: Australian National University, Strategic and Defence Studies Centre, 1974); and Coral Bell, 'Australia in a World of Powers', in P.J. Boyce and J.R. Angel, eds, *Independence and Alliance: Australia in World Affairs 1975–1980* (Sydney: George Allen & Unwin, 1983), 25–37.

Not least this was due to the new Prime Minister's firm views on the Cold War, the importance of the US alliance and the subsequent implications for Australian security and defence policy. Where Whitlam and his government saw an emerging and largely benign, multi-polar order and a superpower confrontation increasingly mediated by détente, Fraser saw continued Soviet belligerence, aided and abetted by a weakening of Western resolve. In this the new Australian Prime Minister was not alone, with an already significant debate in the United States (and amongst its Western alliance partners) as to the benefits of détente. Fraser, as would emerge throughout his tenure, was arguably closer to the more trenchant critics of the Nixon, Ford and Carter administrations, who asserted that the Soviet Union had simply taken advantage of Washington's conciliatory position to pursue military expansion and strategic adventurism, particularly in the developing world. Fraser's guiding assumption with respect to the global balance was clear: T.B. Millar wrote in 1977 that 'He sees the Soviet Union as an unsatisfied power seeking whatever worlds it may conquer short of provoking war with the United States.'[4] In terms of Australian foreign policy this translated into three broad themes: vigorous support of the United States as a global power; efforts to counter perceived Soviet expansionism; and an appreciation of China's anti-Soviet, counter-balancing role in Asia.[5]

When the interdepartmental Defence Committee endorsed an updated *Strategic Basis of Australian Defence Policy* in 1975, the new government had an early opportunity to formulate its views of the global nuclear balance and its implications for Australian security. The Defence Committee's document acknowledged the potential for regional instability, but saw the overall nuclear balance between the superpowers as relatively stable, and emphasized Australia's geographic isolation from the 'primary foci of interest and confrontation between the major military powers'.[6] Fraser himself was clearly not content with the analysis in the paper and asked the Secretary of Defence, Arthur Tange, to review the paper's assumptions in light of the change of government. In doing so, he made particular reference to two paragraphs that implied that Australia 'should be sensitive to Soviet interests' in its cooperation with the United States.[7]

When the Minister for Defence James Killen submitted the paper for consideration by Cabinet in April 1976, he did so without a recommendation for endorsement.[8] That the Cabinet was more focused on the central nuclear

4 Thomas B. Millar, 'From Whitlam to Fraser', *Foreign Affairs* 55, no. 4 (1977): 861.

5 See Girling, 'Australia and Southeast Asia in the Global Balance'; Coral Bell, 'Mr. Fraser and Australian Foreign Policy', *The World Today* 35, no. 10 (1979): 414–20.

6 Defence Committee, *Strategic Basis of Australian Defence Policy* (Canberra: Australian Department of Defence, 3 October 1975), para. 153.

7 Letter from D.J. Killen to Prime Minister Malcolm Fraser, 5 April 1976, NAA A10756, LC183 PART 1.

8 Cabinet Decision No. 448 (FAD), 13 April 1976, NAA A12934, FAD14.

balance than its advisors in the Defence Committee is apparent from the other documents that were also tabled on the same day: Killen submitted four short papers by the Joint Intelligence Organization on *The strategic nuclear balance between the U.S. and the U.S.S.R.*, *Comparison of the balance of forces between China and the U.S.S.R.*, *The growth of Soviet naval power in relation to that of the U.S.* and *NATO/Warsaw Pact military balance in Europe.*[9] Foreign Minister Andrew Peacock tabled a paper summarizing comments by US Secretary of Defense Donald Rumsfeld on the nuclear balance between the United States and Soviet Union.[10] Cabinet did not accept the *Strategic Basis* paper and requested a review 'with particular attention to a full analysis of the global situation, including relations between the great powers and their likely effect on a "neighbourhood" defence policy concept'.[11]

In his June 1976 speech to the House of Representatives on 'Australia and the World Situation', Fraser elaborated on his government's views of the superpower balance. He explained that

> The international diplomacy of the major powers – with which Australia has to deal – has to be understood principally as an effort by these powers to create a balance in the world favourable to their interests. It is in the pursuit of a more favourable balance that their policies impinge on middle Powers, such as Australia, and on areas of immediate importance to Australia, such as South East Asia, the South Pacific area, and the Indian Ocean.[12]

He went on to draw the link between the central nuclear balance and Soviet expansion, stating that

> The U.S.S.R.'s nuclear armory has been elaborated to a point where there is considerable debate about the possibility of the Soviet Union gaining a strategic advantage. The Soviet leaders now have a strategic and political reach – a capacity to influence and even to intervene – well beyond the periphery of the established zones of Soviet security interest.[13]

Rather than the control of a nuclear arms race as such, Fraser thus saw growing Soviet nuclear (and conventional) power as the core problem to international security. 'While the NATO powers' capability remains relatively static, why is the Soviet arms build-up proceeding apace?', he asked, concluding that 'The world cannot afford any reduction of the credibility of the U.S. foreign policy,'

---

9 Cabinet Decision No. 449 (FAD), 13 April 1976, NAA A12934, FAD15.

10 Cabinet Decision No. 450 (FAD), 13 April 1976, NAA A12934, FAD16.

11 Cabinet Decision No. 448 (FAD), 13 April 1976, NAA A13075, 448/FAD AMENDED.

12 *C.P.D.* (Representatives), 1 June 1976, 2736.

13 Ibid., 2737.

and that 'Negotiation is not a substitute for, it is an essential concomitant of, a stable military balance.'[14]

When later in the year the Cabinet had before it a revised *Strategic Basis* paper, entitled *Australian Strategic Analysis and Defence Policy Objectives* (ASADPO), it continued to focus on the importance of the nuclear central balance. An accompanying paper on the *International Security Outlook* was criticized in the Cabinet Decision because 'reference to the SALT talks did not give credence to the fact that the USSR had exploited the agreement to build up its missile inventory'.[15] ASADPO itself was endorsed by Cabinet 'for planning purposes'.[16] It discussed in much greater detail than the earlier version the opportunities that the Soviet Union gained from strategic nuclear parity to exert regional influence, and the importance of Western unity forestalling these.[17]

Hence, it is clear that it was not merely the Soviet Union itself that loomed large in the Fraser government's thinking on security, but specifically the extent to which the central nuclear balance was creating opportunities for Soviet expansion and, thereby, direct or indirect harm to Australia's own security. In conceptual parlance, this was an instance of the 'stability-instability paradox', where it was the relative parity at the nuclear level that gave the dissatisfied power the strategic space of manoeuvre to expand its influence below the level at which the West would contemplate escalation. In practice Australia had of course little scope materially to influence the situation that so occupied the Cabinet's deliberations on the central balance. They would have, however, strengthened the resolve of the government to continue with Australia's support to its ally's nuclear program and capabilities insofar as it could, despite (or because of) its geographic location. In practice, this meant the continued maintenance of the 'Joint Facilities', including but not limited to the submarine communications station at North West Cape, the intelligence satellite facility at Pine Gap, the early warning satellite station at Nurrungar and seismic monitoring facilities.

In 1978, it became publicly apparent that the United States had not informed Australia of the planned installation of a new satellite antenna at North West Cape – leading Defence Minister Killen to comment that the United States was not showing 'the proper courtesy'.[18] However, the government was

14 Ibid., 2737–8.

15 Cabinet Decision No. 6155 (FAD), 1 January 1970, NAA A12909, 696.

16 Ibid.

17 Defence Committee, *Australian Strategic Analysis and Defence Policy Objectives* (Canberra: Australian Department of Defence 1976), paras 30, 351.

18 Quoted in Desmond Ball, *A Suitable Piece of Real Estate* (Sydney: Hale & Iremonger, 1980), 57. The Fraser Cabinet itself 'noted that the US Government had not properly consulted the Australian Government ... and had shown a lack of regard for the relationship between the two countries by not doing so', and decided that the Foreign Minister would discuss the issue with the US ambassador; Cabinet Decision No. 5367, 16 May 1978, NAA A13075, 5367.

neither ignorant, nor naïve or unconcerned about the implications of hosting the facilities. The possibility of them being attacked with nuclear weapons in the – unlikely – event of a superpower war was generally acknowledged in the *Strategic Basis* papers. Fraser himself requested a detailed study in 1980, following the Soviet invasion of Afghanistan, of the effects on Australia of nuclear war in the Northern hemisphere,[19] which was subsequently circulated among the small circle of members of the Cabinet's Foreign Affairs and Defence committee.[20]

Australia continued to assist in the US strategic nuclear program because it saw the benefits of US global deterrence for its security outweighing the risks. When the United States approached Fraser in June 1981 about the possible conduct of extended range tests of the new MX ICBM that would see missiles fired from California into waters off Tasmania, Cabinet saw 'no objection in principle'. However, it also stated that 'there would be a need for close attention to be given to the question of public presentation'.[21] When the Cabinet later returned to the issue, it was clear that the Fraser government had sought a delay of the test until after the 1983 election, and sought to schedule the visit of a US range instrumentation ship outside the expected October/November election period in 1983.[22]

Overall, the Fraser government's policy of providing material support to the US nuclear deterrent, albeit behind a cloak of relative secrecy, continued the policy of its predecessors, including that of Gough Whitlam in government. What was slowly changing by the 1980s was not that the policy itself had become more politically delicate, but that increasingly detailed information about it was becoming available to the Australian public. In particular, ANU Professor Desmond Ball's 1980 book on the joint Facilities, *A Suitable Piece of Real Estate*, injected a wealth of information about Australia's long-standing cooperation into the public debate that, as Richard Tanter would later write, was 'profoundly shocking' to 'a certain generation of Australians'.[23] In the context of the growing peace movement opposed to the Reagan defence build-up, maintaining the secrecy around these facilities and the Australian policy on them became increasingly difficult. In the end, it would be the new Hawke government that would publicly acknowledge and explain what had also been

19 NAA file, 'Policy Paper on the Effects of a Nuclear War on Australia', NAA A10756, LC5130.

20 NAA file, 'Effects on Australia of a Nuclear War – ONA report', NAA A10756, LC3238.

21 Cabinet Decision No. 15591 (FAD), 16 June 1981, NAA A13075, 15591/FAD.

22 Cabinet Submission No. 5312, 2 February 1982, NAA A12909, 5312.

23 Richard Tanter, 'American Bases in Australia Revisited', in Brendan Taylor, Nicholas Farrelly and Sheryn Lee, eds, *Insurgent Intellectual: Essays in Honour of Professor Desmond Ball* (Singapore: ISEAS Publishing, 2012), 193.

the policy of the Fraser government, albeit with new nuances that reflected the changed strategic and political situation of the 1980s.[24]

Non-Proliferation and Australian Security

Although the central nuclear balance was seen by the Fraser government to have significant importance, its effect on Australia's security was ultimately indirect. In the *Australian Strategic Analysis and Defence Policy Objectives* paper (ASADPO), the Defence Committee had therefore recommended that Australia's defence effort should be focused on its own region.[25] This was also the core message of the public 1976 White Paper on *Australian Defence*. It would remain the Fraser government's policy throughout its whole time in office, even after the Carter administration invited Australia to participate in the Rapid Deployment Force for the Persian Gulf in 1980.[26] Regional conflicts would be the most likely threat to Australia.

In this context, however, nuclear proliferation was also a particular concern. In the 1975 Strategic Basis, the Defence Committee pointed out that proliferation would increase the severity of such regional conflicts, and also make them more likely because regional nuclear arsenals would restrict the ability of the superpowers to intervene.[27] In ASADPO, it stated that

> Any further proliferation appears likely to be essentially regional in its motivation and strategic significance, although there would be complex implications for central global relationships were proliferation to become extensive or lead to actual use of nuclear weapons.[28]

---

24 In that sense, the public revelations on the joint facilities arguably helped the long-term political sustainability of the policy itself. Graeme Dobbell writes that: 'The knowledge Ball put into the public realm made it possible for the Australian Labor Party to embrace the US bases and re-commit to the alliance. Australia was able to move to a position of "full knowledge and consent" on the bases, and Des Ball takes much of the credit for killing off a culture of secrecy which had seen the bases as taboo, dark places subject to limited Australian knowledge and even less consent.' Graeme Dobbell, 'Des Ball: An Australian Original', *Lowy Interpreter Blog*, 17 December 2013, available at http://www.lowyinterpreter.org/post/2012/12/17/Des-Ball-An-Australian-original.aspx [accessed 7 February 2013].

25 Defence Committee, *Australian Strategic Analysis and Defence Policy Objectives*, paras 36, 360, 306–11.

26 Cabinet Decision No. 12644 (FAD), 26 August 1980, NAA A12909, 4317.

27 Defence Committee, *Strategic Basis of Australian Defence Policy*, 3 October 1975, paras 37, 38.

28 Defence Committee, *Australian Strategic Analysis and Defence Policy Objectives*, para. 94.

India, Pakistan, Japan, Korea, Taiwan and Indonesia were mentioned as candidates for potential proliferation.[29] The Defence Committee found that 'there is the prospect of the action of one stimulating another',[30] that 'improved safeguards could increase' restraints on proliferation by increasing its technical difficulty,[31] and that '[l]ikely significant reduction in Indonesia's or Japan's lead time to nuclear weapons production would be of particular interest to Australia'.[32] Hence, it concluded that

> It is in the Australian interest that as many countries as possible ratify the Non-Proliferation Treaty. It is also in the Australian interest to support efforts to strengthen and extend international safeguards arrangements governing the supply of nuclear materials, equipment and technology to non-parties as well as parties to the Treaty.[33]

In a way, this position was unremarkable, since it largely reflected the consensus of the proponents of the NPT in Australia during the debates of earlier years. In Australia, the arguments of both proponents and opponents of the NPT had focused on the role of nuclear proliferation in Australia's region – proponents arguing that Australia had much to lose from nuclear countries emerging in its neighbourhood, and opponents pointing out that Australia might well require nuclear weapons itself in situations unrelated to the Cold War.[34]

And yet, this continuity, and the genuine concern with the strategic consequences of regional nuclear proliferation it demonstrates, was also quite remarkable because it was not shared by many other countries at the time. For Northern hemisphere countries during the 1970s, the NPT was primarily a way to manage the Cold War. The NPT negotiations themselves between the United States and Soviet Union assumed significance as a symbol of détente in the 1960s. The superpowers had failed to prevent France and China from developing nuclear arsenals, and for both their (former) allies became the cause of unhelpful strategic complications. For the Soviet Union, nuclear proliferation threatened its status as leader of the Communist world; for the United States,

---

29 In 1980, when Indonesia planned the construction of a nuclear reactor, the Joint Intelligence Organization was 'tasked to give it high priority and monitor any evidence that it might be a covert effort to attain a nuclear weapons capability or shorten the lead time to acquire one'. Paul Dibb, pers. comm., 20 February 2014.

30 Defence Committee, *Strategic Basis of Australian Defence Policy*, 3 October 1975, paras 39, 40.

31 Ibid., para. 36.

32 Ibid., para. 172.

33 Ibid., para. 171.

34 Christopher Hubbard, 'From Ambivalence to Influence: Australia and the Negotiation of the 1968 Nuclear Non-Proliferation Treaty', *Australian Journal of Politics and History* 50, no. 4 (2004): 526–43.

nuclear allies presented a serious challenge to Washington's ability to manage and control nuclear escalation under the strategy of Flexible Response.

Major Western non-nuclear states, especially Japan and West Germany, saw the treaty in the context of their relationship with the United States and the Cold War more generally, rather than because they saw any direct threat or strategic disadvantage from further proliferation. In the German case, ratification in 1975 under the new left-liberal Brandt government was part of its 'Ostpolitik' of reassurance of the Soviet Union – hence, the treaty was seen by both opponents and proponents primarily through the lens of West Germany's delicate position in Europe, rather than as a means to build a global nuclear order.[35] Japan, West Germany and Italy had to negotiate their security during the Cold War among three or four nuclear powers – including allies as well as adversaries. Nuclear proliferation outside the Cold War context mattered little for their security concerns, which were squarely focused on the military forces of the Soviet bloc.

In contrast, Australia and New Zealand were the only Western countries that did not already have a nuclear power in their neighbourhood – and the two that were most close to them, China and India, were also the most recent cases of proliferation. The countries in their neighbourhood were often members of the Non-Aligned Movement (NAM), and the Fraser government was particularly sensitive of the need to build bridges between the West and the 'third world', including on the question of nuclear non-proliferation. Hence, it would prove to be far more willing to support the development of strict safeguards and the NPT regime with its inherent 'bargains' than was most European NATO members – and at times even the United States. Reconciling these interests in non-proliferation with the need for Western unity in the Cold War, and the realization of the economic potential of Australia's uranium, would however require a constant and at times delicate balancing of interests.

Balancing Non-Proliferation and Cold War: The CTB Negotiations

Symptomatic of the tenuous nature of the fledgling nuclear non-proliferation regime, the 1975 NPT Review Conference (RevCon) had been characterized by significant disagreement between member states over key articles of the treaty pertaining to nuclear supply arrangements and disarmament. Only due to the active intervention of the conference president were states able to agree on a final declaration.[36] As Suter subsequently observed, given the intractable nature

35 See Robert M. Lawrence and Joel Larus, eds, *Nuclear Proliferation Phase II* (Lawrence: University Press of Kansas, 1974).

36 See 'The Implementation of Agreements Related to Disarmament', in Stockholm International Peace Research Institute, *World Armaments and Disarmament: SIPRI Yearbook 1976* (Stockholm: SIPRI, 1976), 363–92.

of the differences separating NPT parties at the 1975 RevCon, it would be equally, if not more, difficult to achieve a consensus document in 1980 if these differences were not properly addressed during the five year interlude.[37]

In this context, the Fraser government supported the negotiation of a multilateral Comprehensive Test Ban (CTB) agreement as a core element of strengthening the non-proliferation regime. In his address to the UN General Assembly in 1976, Foreign Minister Andrew Peacock explicitly linked a multilateral CTB agreement with improving the international prospects for nuclear non-proliferation.[38] The same year, Australia registered its support for the annual trans-Tasman Resolution, urging, *inter alia*, that the nuclear powers suspend their respective testing programs 'as an interim step' towards the formal conclusion of a CTB agreement.[39] In 1977, CTB negotiations then began in earnest between the Soviet Union, United States and Britain, after President Carter had approached his Soviet counterpart on the subject.[40] In a speech to the UN Special Session on Disarmament in June 1978, Prime Minister Fraser observed that

> Progress must be made in three inter-related areas. First, the nuclear weapons powers must take effective action to limit and reduce their nuclear arsenals. Second, there must be an end to nuclear testing in all environments. Third, the international non-proliferation system must be strengthened. We are dealing here with difficult and inter-related balances of national interest. Progress in one area can be negated if other essential areas are neglected.[41]

In January 1979, Australia then took its seat as a permanent member of the newly instituted Conference on Disarmament (CD) in Geneva, which gave the Fraser government a new avenue to influence the international agenda on disarmament. In this context, it is clear that Australia saw a very strong link between the CTB and non-proliferation. In a diplomatic cable to several overseas posts in March 1979, Canberra surmised that a CTB would

---

37 Keith D. Suter, 'The 1975 Review Conference of the Nuclear Non-Proliferation Treaty', *Australian Outlook* 30, no. 2 (August 1976): 338.

38 'Foreign Minister Addresses UN General Assembly', *Backgrounder* (Statements) no. 59 (1 October 1976): 7.

39 This was adopted as UNGA Resolution 31/66. See Resolutions and Decisions Adopted by the General Assembly During Its Thirty-First Session, Volume I, 21 September–22 December 1976: General Assembly Official Records, Thirty-First Session, Supplement no. 39 (A/31/39) (New York: United Nations, 1977), 33.

40 R.L. Garthoff, *Detente and Confrontation: American-Soviet Relations from Nixon to Reagan* (Washington, DC: The Brookings Institution, 1985), 756.

41 'Address to the Special Session of the General Assembly Devoted to Disarmament: Text of the Prime Minister's Speech at the United Nations in New York on 5 June 1978', *Backgrounder* (Statements), no. 141 (9 June 1978): 4.

> strengthen the non-proliferation regime … in that after the major nuclear powers have agreed to a test ban the breaking of such a ban by any potential proliferator will be more likely than heretofore to attract a strong international reaction, including the possible application of sanctions.[42]

The leader of Australia's CD delegation, Sir James Plimsoll, stated in August 1979 that

> a consequence of having a [CTB] treaty would be that it would prevent or act against horizontal proliferation. Our aim is a multilateral treaty to which all countries of the world would adhere and it would mean that countries which do not have nuclear weapons today would not be able to acquire them. They would be pledging themselves not to have tests which would make the acquisition or the effective testing of them possible. I think it is important to note that states which have not become parties to the Non-Proliferation Treaty could nevertheless become parties to the Comprehensive Test Ban treaty, and in that way they could give assurances that they would not become nuclear weapon states or that they would not try to become nuclear weapon states, even though they felt that they could not become parties to the Nuclear Non-Proliferation Treaty.[43]

Unfortunately for Australia, its accession to the CD coincided with what is generally regarded as the high point of trilateral CTB negotiations. As one observer noted, '[b]y 1980, the trilateral talks had become another ritual negotiation'.[44] At that point, it had also become obvious that important differences separated Australia and the United States on the CTB issue; not least on the question of the length of any treaty. The Fraser government maintained that it be of indefinite duration, while the Carter administration insisted that its maximum duration (at least initially) be confined to three years.[45]

These differences came to the fore during the 1980 NPT RevCon. Constraining the 'vertical' proliferation of the declared nuclear powers had been just as much an objective for many states in negotiating the NPT as had been checking the 'horizontal' spread of nuclear weapons. The disenchantment

---

42 Australian Department of Foreign Affairs, Outward Cablegram, 'Second NPT Review Conference: Prospective Issues', O.CH814185 1615, 23 March 1979, NAA File: 919/10/5/2, no. 8, 2.

43 'Comprehensive Nuclear Test Ban: Statement by Sir James Plimsoll – 2 August 1979', in *Report to Parliament on the 1979 Session of the Committee on Disarmament, Geneva, 24 January to 26 April and 14 June to 14 August* (Canberra: Australian Government Publishing Service, 1979), Annex E, 43.

44 Carter, *Success and Failure*, 90.

45 For discussion on this point more generally, see 'Nuclear Explosions', in Stockholm International Peace Research Institute, *World Armaments and Disarmament: SIPRI Yearbook 1982* (London: Taylor and Francis, 1982), 434.

felt by many NNWS over the failure of the NWS to achieve progress in implementing Article VI after 1975 was reflected in analyses shortly before the 1980 Conference, some of which canvassed the possibility of withdrawals from the Treaty in the event of an inconclusive Conference outcome.[46] High-ranking Foreign Affairs officials held fears that the Conference would collapse due to disillusionment among NNWS, particularly those comprising the NAM. This apprehension was evident in correspondence from a senior official of the Department's Nuclear Policy Branch to the First Secretary at the Australian Mission to the United Nations:

> Australia should be very concerned to avoid any erosion of support for the Treaty. Our approach is, therefore, partly defensive. We expect that the major debate will be about access, or lack of it, to nuclear technology, particularly by developing states … [t]he best we can hope for is that disagreements are sufficiently contained so that achievement of consensus at the NPTRC will be possible. If tangible preferences for NPT States can be demonstrated *AND* significant progress on Article VI issues achieved by the time of the RC, this should be possible. Even under those circumstances, avoidance of a breakdown may be a close shave.[47]

During the Preparatory Committee (Prep Com) phase before the 1980 Conference, the Fraser government's intended strategy and approach began to take shape. Reflecting Australia's particular interest in the viability of the NPT among the NAM, the Department of Foreign Affairs emphasized that Australia would not be locked into supporting either US policy or the aims of the Western Group[48] as a whole. In a briefing for the Australian delegation to the first Prep Com meeting, the Department of Foreign Affairs noted that:

> On nuclear disarmament issues we have views which are close to non-aligned states, in particular Mexico, Yugoslavia and Nigeria in addition to the previously mentioned states [Canada, the Netherlands, Norway and Sweden]. On questions of access to nuclear technology and safeguards, we are closer to the USA and Canada among nuclear supplier states. As an aligned state we take into account

---

46 See for instance Jack Barkenbus, 'Whither the Treaty?', *Bulletin of the Atomic Scientists* 36, no. 4 (April 1980): 37–9; and 'The Anti-Bomb-Spread Treaty is Still Worth Saving', *The Economist*, 16 August 1980, 27–8.

47 Australian Department of Foreign Affairs, 'Letter to Mr Allan Behm, First Secretary, Australian Mission to the United Nations, Geneva, from J.A. Benson, Acting Assistant Secretary, Nuclear Policy Branch, 4 April 1979', NAA File: 919/10/5/2, Part: 8, 1. Emphasis in original.

48 Similar to the UN General Assembly, CD and later RevCons, individual NPT member states were part of one of four separate Groups at the 1980 NPT RevCon: the Western, Socialist, NAM and the smaller Neutral Group.

> security interests which arise in the context of our own immediate defence Treaty obligations or generally in the interests of the Western aligned systems. *However, we do not regard non-proliferation as an issue about which there should be a common Western aligned view on every aspect.* Consultations with other delegations should accordingly [be] set against this background.[49]

During the General Debate it became clear that many NAM states had come to the Conference determined to insist upon the CD adopting a mandate to negotiate a CTB treaty as evidence of the good faith of the NWS in relation to Article VI.[50] At a relatively early stage, the Australian delegation advised Canberra that divisions over the CTB issue alone could nullify any chance of agreement on a final draft declaration.[51] Australian diplomatic communications during the RevCon displayed frustration with the refusal of the United States to address the concerns of NAM states regarding the CTB issue. Surveying American resistance to suggestions by some NAM parties that basic points on a CTB treaty be incorporated in the final declaration, in late August the Australian delegation noted that Washington's 'stone-walling tactics lack credibility and complicate the task of working for an outcome satisfactory to all'.[52] Responding to calls from Washington for 'Western solidarity' in the final days of Conference proceedings, Australian officials sharply observed that

> the United States might consider the important interests which its friends and allies have at stake in this Conference as well as the likely adverse effects on the [nuclear non-proliferation] regime of failure to reach consensus.[53]

In the end, the 1980 NPT RevCon failed to agree on a final declaration. Washington's aversion to countenance any role for the CD in negotiating a treaty seemed to confirm that non-proliferation concerns had effectively been marginalized in American CTB policy. Writing in 1980, Towle observed that Australia had 'pushed harder for a Comprehensive Test Ban Treaty than the

---

49 Australian Department of Foreign Affairs, 'Brief for the First Meeting of the Preparatory Committee Meeting for the NPT Review Conference – Geneva, 17–20 April', 23 March 1979, NAA File: 919/10/5/2, Part: 9, 10. Emphasis added.

50 Australian Department of Foreign Affairs, Inward Cablegram, 'NPTRC: CTB', O.GE46275 1900, 20 August 1980, NAA File: 919/10/5/2, Part: 16, 1.

51 Australian Department of Foreign Affairs, Inward Cablegram, 'NPTRC: CTB Issue', O.CH910602 1630, 21 August 1980, NAA File: 919/10/5/2, Part: 17, 1.

52 Australian Department of Foreign Affairs, Inward Cablegram, 'Comprehensive Test Ban: NPT Review Conference', O.GE46441 1830, 27 August 1980, NAA File: 919/10/5/2, Part: 17, 1.

53 Australian Department of Foreign Affairs, Inward Cablegram, 'NPTRC: United States Attitude', O.GE46529 2000, 1 September 1980, NAA File: 919/10/5/2, Part: 17, 1.

British and Americans would like'.[54] According to Reus-Smit, while the Fraser government avoided supporting arms control initiatives which the United States opposed outright (such as proposals for a nuclear freeze), it 'continued to back the CTB in the face of growing American ambivalence after 1979'.[55] In October 1982, Foreign Minister Tony Street praised his government's performance, noting that it had been 'a leading proponent within both the United Nations and the Committee on Disarmament of the early negotiation of a comprehensive nuclear test ban treaty'.[56]

And yet, this claim was far easier to sustain for the period until 1980 than the years that followed. The Fraser government continued to avow support in UN General Assembly resolutions for the conclusion of a CTB treaty in general, and a role for the CD in the negotiations particular. It did not, however, pursue this line within the CD itself.[57] Although the Carter administration had suspended the CTB negotiations in 1980 – a decision that the Reagan administration showed no intention to reverse – in July 1981 Foreign Minister Street declared that American nuclear non-proliferation policy was '*fully* consistent with [that] of Australia'.[58]

Until 1980, Australia's policy had been to leverage the CTB negotiations to gain broader support, especially among the NAM, for the NPT regime. This had, however, failed to prevent the failure of the 1980 RevCon, and prospects for a successful conclusion of a CTB agreement rapidly diminished thereafter. Approaching the CTB primarily through a non-proliferation lens in the years that followed would have brought Australia into direct conflict with US policy, with little prospect of changing basic attitudes of the NAM. Moreover, the international context for arms control had been radically transformed by the 'dual-track' decision of NATO member countries in December 1979 to introduce nuclear intermediate-range forces to Europe by 1983, should the Soviet Union not agree to limit its equivalent forces. This prompted Soviet diplomats (and a large 'peace' movement in northern Europe) to place heavy pressure on NATO governments to reverse this decision,[59] and corresponding

54 P. Towle, 'Australian Policy in the Committee on Disarmament', *Strategic and Defence Studies Centre Working Paper*, no. 18 (Canberra: Strategic and Defence Studies Centre, Australian National University, 1980), 6–7.

55 C. Reus-Smit, 'Arms Control, Nuclear Strategy and Australian Foreign Policy: The Fraser Years', *Interdisciplinary Peace Research* 1, no. 2 (October/November 1989): 68.

56 *C.P.D.* (Representatives), 28 October 1982, 2707.

57 Andrew O'Neil, 'Constraints and Continuity: The Evolution of Australia's Non-proliferation Policy, 1975–96' (unpublished Doctor of Philosophy Thesis, Flinders University, 1999).

58 'Nuclear Non-Proliferation', *Commonwealth Record* 6 (13–19 July 1981), 833. Emphasis added.

59 See Thomas Risse-Kappen, 'Did "Peace Through Strength" End the Cold War? Lessons from INF', *International Security* 16, no. 1 (Summer 1991): 162–88.

concern in Western capitals about maintaining unity on nuclear weapons policy. Hence, in 1980 Australia was still sympathetic that the CD become involved in the CTB negotiating process, while the United States, Soviet Union and Britain were more reluctant.[60] When the NAM states were joined by the Socialist group in the CD's 1981 session in advocating the creation of a Working Group to negotiate a treaty text, the Australian delegation however refused to support this call.[61] By the following year, Australian statements clearly attributed the deterioration in the international climate for nuclear arms control, including the CTB, to Soviet actions in Europe and the Third World.[62]

The 'Plutonium Economy': Nuclear Energy and Proliferation in the 1970s

Fraser and his Liberal and Country Party colleagues had in Opposition been vocal advocates for the immediate development of Australia's uranium for export, and viewed the Ranger Inquiry as an obstacle to this objective. Thus when the Coalition won the December 1975 federal election it was assumed by many that uranium mining and export would be rapidly approved. It was clear that the new Fraser government not only wished to grasp the economic opportunities presented by the deposits but also, like its predecessors, to explore the potential for developing a uranium enrichment capability.[63]

Yet its ability to do so was critically influenced by shifts in the economic and normative domains of nuclear policy at the international level. Most important in this regard were the initiatives of the Ford and Carter administrations that sought to control the diffusion of nuclear technology, in an effort to curb further horizontal nuclear proliferation. In particular, the focus of the Carter

60 *Report of the Committee on Disarmament, 1980: General Assembly Official Records, Thirty-Fifth Session, Supplement*, no. 27 (A/35/27) (New York: United Nations, 1980), 12–13.

61 United Nations, *Yearbook of the United Nations: 1981* 35 (New York: United Nations Department of Public Information, 1985), 64; see for instance Australian Department of Foreign Affairs, 'Australian Statement of 5 February 1981', Annex C, and 'Australian Statement of 24 April 1981', Annex D, in *Report of the Australian Delegation to the 1981 Session of the Committee on Disarmament* (Geneva, Switzerland, 1981), 38–41 and 2–45.

62 See 'Statement by Mr David Sadleir, Ambassador, Permanent Representative of Australia, 11 February 1982', in *Report of the Australian Delegation to the 1982 Session of the Committee on Disarmament, Geneva, 2 February to 17 September 1982* (Canberra: Australian Government Publishing Service, 1982), Annex D.

63 See for example Cabinet Submission No. 63, 30 January 1976, NAA A1209, 63, in which the Fraser government's Minister for National Resources, Doug Anthony, asserted 'we should examine the possibility of establishing a uranium enrichment plant in Australia'.

administration on controlling the spread of both spent fuel reprocessing and uranium enrichment technology raised both challenges and opportunities for the Fraser government. On the one hand, President Carter's desire to limit the number of states with enrichment capabilities presented a roadblock to Australia's long-standing desire to develop an indigenous enrichment capacity to 'value add' to its exports of uranium. On the other hand, a shift away from spent fuel reprocessing within the international nuclear fuel cycle could increase global demand for uranium. The Carter administration's desire to achieve such a shift therefore opened a window of opportunity for the Fraser government to have its yellowcake and to eat it too, by enabling it to frame its desire to export uranium as an important and responsible contribution to global efforts to halt nuclear proliferation and promote energy security.

During the 1970s three developments – the rapid growth in civilian nuclear power generation in the US, Japan and Western Europe, an intensification of international terrorism, and the testing of a nuclear device by India in 1974 – converged to make questions of nuclear proliferation and nuclear energy central policy issues for governments around the globe. A major question raised by these dynamics was whether expanded use of nuclear power to meet increasing energy demand would ultimately increase the likelihood of nuclear weapons proliferation. India's 'peaceful nuclear explosion' (PNE) of May 1974 was taken by many to be a harbinger of things to come in this regard. India's diversion of plutonium, extracted from the spent fuel rods of a Canadian-supplied CIRUS reactor, and heavy-water supplied by the US, brought into stark relief the need for more stringent export controls and safeguards for nuclear materials and technology in order to limit further horizontal proliferation.[64] A key policy choice for the United States, and other major suppliers of nuclear materials and technology with an interest in slowing or halting further nuclear proliferation, was whether to attempt to achieve non-proliferation goals through multilateral cooperation, or through unilateral measures to deny potential proliferators access to nuclear technology (e.g. nuclear reactors, uranium enrichment and reprocessing plants) and materials (i.e. uranium or plutonium).[65]

The responses of the Ford and Carter administrations to this question oscillated between these poles, creating significant dilemmas for both consumers and suppliers of nuclear materials and technology. Throughout the 1970s, the US retained three significant elements of leverage through which it could hope to influence the development of the economic domain of global nuclear affairs:

64 See Paul F. Power, 'The Indo-American Nuclear Controversy', *Asian Survey* 19, no. 6 (1979): 574–96.

65 For a detailed overview of the effect of the PNE on US policy see J. Samuel Walker, 'Nuclear Power and Nonproliferation: The Controversy over Nuclear Exports, 1974–1980', *Diplomatic History* 25, no. 2 (2001): 215–49; see also Frank von Hippel, 'Plutonium and Reprocessing of Spent Nuclear Fuel', *Science* 5539, no. 28 (2001): 2397–8.

its possession of the world's most advanced nuclear technology; its position as a primary source of financing for the construction of nuclear facilities (including reactors); and its near monopoly status as supplier of low enriched uranium fuel in the 'free world'.[66] This dominant position was, however, increasingly challenged by the emergence of new suppliers such as France, West Germany and Japan. These states were expanding their own nuclear power programs and, as their domestic markets did not appear large enough to support such expansion alone, sought to sell not only reactor technology but also the full range of nuclear fuel cycle facilities abroad. Even in the wake of the furore surrounding the Indian PNE, these emerging nuclear suppliers exerted strenuous efforts to secure major export markets. Yet, Washington feared that increased competition amongst nuclear suppliers held the potential to undermine the goal of non-proliferation, as suppliers would be tempted to present potential customers with various 'sweeteners' in the form of relaxed conditions and safeguards in the quest for competitive advantage.[67]

France and West Germany in particular led the way in this regard, concluding expansive nuclear agreements with Brazil, South Korea and Pakistan. The conclusion of a West German agreement with Brazil in February 1975 was particularly symptomatic of the challenges that increased competition posed for US non-proliferation goals. The deal, worth an estimated $8 billion over a 10 to 15-year period, involved the transfer of full fuel cycle technologies – i.e. the export of technology for uranium mining, uranium enrichment, fuel fabrication, reactors and reprocessing – to a country that had signed neither the NPT, nor a safeguards agreement with the IAEA.[68] Washington, stung by the Indian PNE, immediately raised objections with Bonn over the deal, urging the West Germans to reconsider in light of serious proliferation concerns generated by the projected transfer of the full fuel cycle. For West Germany, however, commercial considerations were paramount, and US appeals for greater safeguards were dismissed as hindrance to its competitiveness.[69]

In an effort to come to grips with this issue, the Ford administration convened discussions with the other major nuclear suppliers, the Soviet Union, West Germany, France, Canada, Japan and the United Kingdom in early 1975.

---

66 Some estimates suggest the US supplied 70% of the 'free world's' nuclear fuel and technology in the mid-1970s. See Walker, 'Nuclear Power and Nonproliferation'; Peter Tzeng, 'Nuclear Leverage: US Intervention in Sensitive Technology Transfers in the 1970s', *The Nonproliferation Review* 20, no. 2 (2013): 482–4.

67 See Karl Kaiser, 'The Great Nuclear Debate: German-American Disagreements', *Foreign Policy* 30 (Spring 1978): 83–110; William Glenn Gray, 'Commercial Liberties and Nuclear Anxieties: The US-German Feud over Brazil, 1975–7', *International History Review* 34, no. 3 (2012): 449–74.

68 Tzeng, 'Nuclear Leverage', 474.

69 Steven J. Baker, 'Monopoly or Cartel?', *Foreign Policy* 23 (Summer 1976): 202–20.

The administration's goals in these discussions were fivefold: to strengthen IAEA safeguards and to ensure they were effective and applied; to slow the spread of uranium enrichment and spent fuel reprocessing facilities; to 'impose special conditions' on exports to 'volatile' regions; to enhance the physical security of nuclear materials; and to minimize the risk that nuclear materials or technology from supplier states would be used for further PNEs.[70] The group announced in February 1976 that it had reached agreement on strengthening export controls over nuclear materials and technology that met all but one of the Ford administration's goals: export of uranium enrichment and fuel reprocessing technology. In this regard the other members of the London Suppliers Group only agreed to 'exercise restraint' with respect to the export of such technology.[71]

During the US presidential election campaign of 1976, Democratic candidate Jimmy Carter distinguished himself from the preceding Republican administrations of Nixon and Ford by explicitly linking the issues of domestic nuclear energy use and non-proliferation. Throughout the campaign, Carter drew attention to the national and international security implications for the US of the continued spread of nuclear technology.[72] For Carter, the Indian PNE of 1974 demonstrated the proliferation danger inherent in the further spread of ostensibly civilian nuclear technology and, for this reason, he also publicly called for the US to oppose the West German-Brazil nuclear agreement of 1975 and French attempts to export enrichment and reprocessing technology to Pakistan.[73]

However, the oil crisis of 1973 had also stimulated renewed interest and demand for nuclear energy as a possible route to energy security, with a subsequent projected rise in demand for uranium fuel to power nuclear reactors. This in turn generated questions as to the availability of uranium fuel, prompting fears that a greater reliance on nuclear power would in fact result in the replacement of dependence on one set of energy suppliers (of oil) with dependence upon on another set (of uranium fuel). This fear, it was hoped, could be mitigated by the rapid development and use of plutonium fuel – through the operation of fast 'breeder' reactors – that would permit countries to become self-sufficient in nuclear fuel.[74] However, this projected turn towards

70 Walker, 'Nuclear Power and Nonproliferation', 225.

71 Ibid., 225–6.

72 J. Michael Martinez, 'The Carter Administration and the Evolution of American Nuclear Nonproliferation Policy, 1977–1981', *Journal of Policy History* 14, no. 3 (2002): 264.

73 Lawrence G. Franko, 'US Regulation of the Spread of Nuclear Technologies through Supplier Power: Lever or Boomerang?', *Law and Policy in International Business* 10, no. 4 (1978): 1191–2.

74 P. Andrew Karam, 'How do fast breeder reactors differ from regular nuclear power plants?', *Scientific American*, 17 July 2006, available at http://www.scientificamerican.com/article.cfm?id=how-do-fast-breeder-react [accessed 20 July 2014].

fast breeder reactor technologies as a means of combating energy insecurity held the potential to place nuclear weapons within even closer reach of those countries that operated them, as such reactors were essentially designed as factories for the production of fissile material.[75]

This was a particular concern for the Carter administration with respect to the proposed breeder reactor programs of allies such as Japan, West Germany and Italy as well as those of Third World states such as Pakistan.[76] For the Carter administration the key to mitigating this threat to international security was to control the spread of reprocessing technology or the 'plutonium economy'.[77] The administration's focus on the proliferation potential of breeder reactors overlooked the fact that the major nuclear reactor types then in commercial use – light-water reactors fuelled by enriched uranium and heavy-water reactors fuelled by natural uranium – also produced plutonium as a by-product of their operation.[78] In fact subsequent decades would demonstrate that the proliferation potential of uranium enrichment technology would be equally as dangerous as that associated with the production of plutonium.[79]

The Fraser Government's Uranium Decision

Prime Minister Fraser was quick to grasp the importance of this shifting international context for Australian policy. In a letter to President Carter on 4 February 1977 (subsequently tabled in Federal Parliament on 23 March), he wrote that Australia 'fully' supported the objective of strengthening the

75 For the Carter administration's views see Joseph S. Nye, 'Nonproliferation: A Long-Term Strategy', *Foreign Affairs* 56, no. 3 (1978): 604–6. Nye was then Deputy to the Under Secretary of State for Security Assistance, Science and Technology and chaired the National Security Council Group on Nonproliferation.

76 See T.T. Poulose, 'Nuclear Proliferation: A Third World Perspective', *The Round Table* 69, no. 274 (1979): 156–8; Antonio Tisco, 'The Carter Administration and its Non-Proliferation Policies: the Road to INFCE', *Humana Mente: Journal of Philosophical Studies* 16 (2011), 60–64.

77 Martinez, 'The Carter Administration and the Evolution of American Nuclear Nonproliferation Policy, 1977–1981', 267–8.

78 See for example Mason Wilrich, 'A Workable International Nuclear Energy Regime', *Washington Quarterly* 2, no. 2 (1979): 13–30.

79 The exposure of the proliferation activities of the A.Q. Khan network in 2003–04 and of Iran's nuclear program, for example, would starkly demonstrate the proliferation potential of uranium enrichment technology. See Chaim Braun and Christopher Chyba, 'Proliferation Rings: New Challenges to the Nuclear Non-Proliferation Regime', *International Security* 29, no. 2 (2004): 5–49; Christopher Clary, 'The AQ Khan Network and the Limits of the Non-Proliferation Regime', *Disarmament Forum* 4 (2004), 33–42.

non-proliferation regime and concurred with the President's warnings about the proliferation dangers posed by spent fuel reprocessing and plutonium.[80] Fraser also noted that beyond 'our general commitment to non-proliferation':

> Australia's particular interest – and *perhaps our scope in future to exert influence on international developments – relates to our potential as a supplier of uranium.* Australia would certainly want nuclear material deriving from any uranium it might supply to be subject to stringent control.[81]

Fraser thus perceived that the US non-proliferation initiatives presented Australia with an opportunity to support international non-proliferation efforts through the export of its uranium. Carter responded to Fraser's overture by providing a glimpse of his administration's policy direction, noting that:

> we are actively examining ways to provide guaranteed fuel supplies to countries which are willing to accept constraints consistent with our non-proliferation objectives. This will help to reduce proliferation by giving nations an incentive to place their nuclear facilities under international safeguards and not to acquire sensitive nuclear facilities.[82]

The President also concurred with Fraser's presumption that Australian uranium could play a positive role in this regard, by suggesting that 'Australia's potential as a major supplier of uranium gives you a particular interest in this aspect [i.e. guaranteed fuel supplies] of the subject.'[83] Furthermore, the President argued, if the US, Canada and Australia could 'collaborate on policies for the supply of natural uranium, we can play a vital role in reducing the threat of proliferation'.[84]

The direction of the Carter administration's efforts to prevent the further spread of the 'plutonium economy' was subsequently revealed with the President's announcement of a number of changes in US nuclear policy on 7 April 1977. These included: the indefinite US deferral of commercial reprocessing and recycling of plutonium; halting plans for the liquid metal fast breeder reactor (LMFBR) program previously endorsed by the Nixon and Ford administrations; an increase in US capacity to produce enriched uranium fuels; and the imposition of an embargo on the export of enrichment and reprocessing technology. The President also indicated that the United States would be exploring technical options to minimize the proliferation dangers of

80 *C.P.D.* (Representatives), 23 March 1977, 484.
81 Ibid. Emphasis added.
82 Ibid.
83 Ibid.
84 Ibid.

the nuclear fuel cycle while simultaneously attempting to assure energy deficient states such as Japan of long-term supplies of energy.[85]

Significantly, the President also publicly reiterated his belief that Australia, as a potentially 'substantial' supplier of uranium, had a key role to play in assisting his administration in dissuading other states from adopting the plutonium route, with its attendant proliferation dangers, to nuclear energy.[86] The President's desire to collaborate with Australia (and Canada) on this issue was further expressed through the convening of a series of discussions in April and May 1977 of senior US, Australian and Canadian officials concerning the practicality of various nuclear safeguards proposals.[87] During these talks it was made clear that the Carter administration believed the export of Australian uranium would be 'pivotal' in discouraging the use of plutonium.[88]

President Carter's initiatives to move the nuclear fuel cycle away from plutonium and towards enriched uranium fuels thus opened a window of opportunity for uranium suppliers such as Australia. This was especially fortuitous for the Fraser government as it sought to neutralize the domestic political debate surrounding the two reports issued by the Ranger Environmental Inquiry in October 1976 and May 1977. By framing the export of Australian uranium as a significant contribution to the US quest to halt the 'plutonium economy', it hoped to achieve two central goals: to neutralize concerns expressed in the Ranger reports regarding the potential connection of Australian uranium exports to nuclear proliferation (both horizontal and vertical); and to simultaneously demonstrate that Australia had a major commercial opportunity to become a major supplier of safeguarded uranium.

The government waited to publicly present its approach to these issues until after the release of the Ranger Inquiry's second report in May 1977, which was followed by a number of major statements by the Prime Minister and other senior Ministers on 24 May 1977 and 25 August 1977. The first specified the conditions under which Australia would export uranium if the government chose to do so, thereby establishing the foundations of Australia's nuclear safeguards policy. The second consisted of major statements to parliament by both Prime Minister Fraser and then Acting Minister for Foreign Affairs, Ian Sinclair, on 25 August 1977. These latter statements presented a broader picture

85 'Nuclear Power Policy Remarks and a Question-and-Answer Session with Reporters on Decisions Following a Review of US Policy, 7 April 1977', *Public Papers of the Presidents of the United States: Jimmy Carter, January 20 to June 24, 1977*, *The American Presidency Project*, available at http://www.presidency.ucsb.edu/ws/index.php?pid=7315 [accessed 20 July 2014].

86 Ibid.

87 Australian Atomic Energy Commission, *Australian Atomic Energy Commission Annual Report: 1976–77* (Canberra: Australian Atomic Energy Commission, 1977), 47.

88 Paul Kelly, 'The Carter Connection', *The National Times*, 30 May–4 June 1977; 'Australian Uranium: Anthony Supports Carter', *The Economist*, 21 May 1977, 93.

of the logic behind the government's final decision to approve uranium exports. The major factor shaping both the policy direction adopted in the nuclear safeguards statement of 24 May and the rhetoric of the second statement on uranium export policy of 25 August was the non-proliferation initiatives launched by the Carter administration.

In the presentation of his government's safeguards policy, the Prime Minister began by noting that while the issue of preventing the further spread of nuclear weapons 'without forgoing the tangible benefits of nuclear power' was one of 'major international importance', it had 'added significance for Australia because of our potential as a supplier of uranium'.[89] Fraser then asserted that his government was 'determined that Australia should play an active role with other countries in the search for, and achievement of, joint solutions' to this international problem, and reaffirmed Australia's commitment to non-proliferation, arguing that the 'proliferation of nuclear facilities without adequate protection against diversion of material to nuclear weapons production' was 'obviously inimical to Australia's interests and to global and regional security'.[90] These factors accordingly required 'selectivity in the choice of customer countries and the closest attention to adequate safeguards' and that safeguards not be something 'to be balanced against commercial considerations'.[91]

These considerations were clearly reflected in the seven major elements of nuclear safeguards policy the Prime Minister announced:

- As international efforts to strengthen and improve nuclear safeguards was an ongoing process, the government would keep Australian policy under 'regular review';
- Australia would not export to NNWS not party to the NPT, and recipients had to be members in 'good standing' of the NPT. All NWS would have to provide Australia assurance that nuclear material supplied by it for peaceful purposes not be diverted for military purposes;
- Australian-obligated uranium would be covered by IAEA safeguards from the time it left Australian ownership and nuclear material supplied by Australia would be subject to safeguards for the full life of the material;
- Bilateral safeguards agreements between the Australian government and recipient countries would be required for all future contracts;
- Such bilateral safeguards agreements would be 'full scope', i.e. re-transfer, enrichment or reprocessing of Australian uranium would be contingent on the prior consent of the Australian government;

89 The Rt. Hon. Malcolm Fraser, MP, 'Government Policy on Nuclear Safeguards – Ministerial Statement', *C.P.D.* (Representatives), 24 May 1977, 1700–705. This statement reflected in essential detail the decision of the Fraser Cabinet of 23 May 1977; see Cabinet Decision No. 2988, 23 May 1977, NAA A12909, 2988.

90 Ibid., 1701.

91 Ibid.

- Importing countries would have to provide an assurance that their domestic nuclear facilities had adequate physical security; and
- Australia would actively seek to coordinate nuclear safeguards with other 'like minded' states.[92]

Fraser in conclusion argued that the 'essential ingredients' of his government's policy were 'careful selection of customer countries', 'the application of international safeguards', 'additional safeguards through bilateral agreements' and 'active involvement by Australia in international efforts' to strengthen nuclear safeguards.[93] That this dovetailed in essential detail with the goals of the Carter administration's non-proliferation policies was not lost on the government's critics. Paul Keating, responding for the Labor opposition, trenchantly observed that the government's safeguards policy was so close to that of the United States that it amounted to 'a "Cart it to Carter" policy' driven by Fraser's desire to 'shore up what he sees as an opportunity with the American President'.[94]

The impression, however, that Australia did in fact have an opportunity to establish itself as a key partner of the US administration on the issues of uranium and non-proliferation was further underlined during the Prime Minister's subsequent visit to the United States in June, during which President Carter asserted that joint US and Australian efforts to prevent further proliferation was 'the most important new concept that binds us'.[95] Australia's emergence as a significant and 'responsible' uranium exporter would help, it was hoped, give effect to the administration's desire to reduce the spread of the 'plutonium economy' by, in the President's words, providing 'guaranteed fuel supplies to countries which are willing to accept constraints consistent with our non-proliferation objectives'.[96]

These goals underpinned the justifications provided by the Fraser government on 25 August for its decision to approve the export of uranium. The Prime Minister declared that after 'exhaustive consideration' and 'motivated by a high sense of moral responsibility' his government had decided that 'there should be further development of uranium under strictly controlled conditions'.[97] This position was determined by the government's recognition that nuclear proliferation was the most serious potential danger stemming from a projected expansion in demand for nuclear energy and uranium. The response to these

92 Ibid., 1702–4.

93 Ibid., 1705.

94 Rt. Hon. Paul J. Keating, MP, *C.P.D.* (Representatives), 24 May 1977, 1705–7.

95 Cited in Henry S. Albinski, 'Problems in Australian Foreign Policy – July to December 1977', *Australian Journal of Politics of History* 24, no. 1 (April 1978): 18.

96 See President Carter's 11 March 1977 reply to Fraser in *C.P.D.* (Representatives), 23 March 1977, 484.

97 The Rt. Hon. Malcolm Fraser, MP, 'Australian Uranium Policy – Ministerial Statement', *C.P.D.* (Representatives), 25 August 1977, 645.

twin challenges 'was best served by Australia agreeing now to the export of uranium'.[98] In a situation in which 'Australia possesses 20 per cent of the world's known reserves of low cost uranium', Fraser continued, 'we are in a special position of influence and have a corresponding moral responsibility to maximise protection against nuclear weapons proliferation by responding to the needs of many countries for adequate assurances of uranium supplies'.[99] Echoing the logic of the Carter administration's approach, Fraser argued that the export of Australian uranium could 'slow the movement towards the use of plutonium as a nuclear fuel and lessen the attendant risks of nuclear weapons proliferation'. This stance would also 'greatly' strengthen Australia's ability 'to support more effective safeguards and minimise proliferation risks', as '[o]nly as a producer and supplier of uranium can Australia be an effective force'.[100] The Prime Minister concluded by drawing together these arguments in a formulation that would in time become a bipartisan justification for the export of Australian uranium:

> The export of Australian uranium will decrease the risks of further proliferation of nuclear weapons and will support and strengthen the Nuclear Non-Proliferation Treaty. It will help to make a safer world. The advent of Australia as a major supplier of uranium will make certain that Australia's voice on this most vital problem of international affairs – nuclear weapons proliferation – will be heard and will be heard with effect.[101]

The Acting Minister for Foreign Affairs, Ian Sinclair, further underlined the extent of the government's embrace of the central element of the Carter administration's approach. Sinclair reiterated the four 'cornerstones' of the government's nuclear safeguards policy – 'strict' adherence to the NPT, IAEA safeguards, bilateral agreements between Australia and importing countries and Australian participation in multilateral efforts to strengthen safeguards and the non-proliferation regime – that had been presented to parliament by the Prime Minister on 24 May.[102] He provided further explanations to the parliament about the linkages between aspects of nuclear safeguards policy and

98 Ibid., 646.

99 Ibid., 647. That Australia had an obligation to supply raw materials was an enduring theme in Fraser's public statements. In March 1975, for example, he argued that Australia not only had 'an obligation to use our natural resources … for the wider good of the international community' but also 'to be an assured and honest supplier of raw materials'. See Malcolm Fraser, 'National Objectives – Social, Economic, and Political Goals', *Australian Quarterly* 47, no. 1 (1975): 33.

100 *C.P.D.* (Representatives), 25 August 1977, 647. Emphasis added.

101 Ibid., 651.

102 The Rt. Hon. Ian Sinclair, MP, 'Uranium Development: International Implications – Ministerial Statement', *C.P.D.* (Representatives), 25 August 1977, 660.

the government's justification for approving the export of uranium. For the government, the NPT was the 'most important international non-proliferation instrument' and its uranium export policy was designed to further the goal of achieving 'universal adherence' to the Treaty. The government's uranium export policy would make a direct contribution here by offering 'a tangible reward – namely access to Australian uranium' for those countries which were prepared to renounce nuclear weapons by signing the NPT. This 'tangible incentive' was, in the government's estimation, also capable of encouraging adherence to the Treaty by 'countries not at present party to it'.[103]

Ultimately the government's policy was based on the recognition that it was 'not realistic to ask countries to renounce nuclear weapons and accept stringent safeguards while denying them the means to develop nuclear energy for peaceful purposes'.[104] Furthermore, the issue of 'assured supplies of natural and low enriched uranium' was 'doubly important at the present time, because inadequacy of such supplies would reinforce the trend towards the use of technologies based on plutonium'.[105] Given this context, Sinclair asserted, it would be 'a fundamental error to suppose that uranium export and the objective of non-proliferation are incompatible'.[106] However, the only way to that the government's 'stringent safeguards policy' would have an effect and be 'incorporated in new international arrangements', Sinclair underlined, would be if Australia could 'speak from a position of strength'. Such a position could only be obtained, and its nuclear safeguards policy made operative, if Australia was 'seen as a country which has legitimate and direct interests and which is able to offer a tangible benefit in return for acceptance by others of these stringent controls'. In the government's view, uranium exports subject to effective safeguards, 'Far from hindering the cause of non-proliferation … will place Australia in a position to help the development of an increasingly effective non-proliferation regime'.[107]

The embrace of this position made sense for the government for three major reasons. First, Australia, by agreeing to the export of uranium, was demonstrating its commitment to fulfilling its obligations in relation to facilitating the civil use of nuclear energy under the NPT. Second, as demonstrated by Prime Minister Fraser's arguments in his correspondence with President Carter and his statements to federal parliament, the government perceived that the export of uranium in this context would create a position of influence for Australia not only within its relationship with the United States but also in multilateral non-proliferation efforts. Finally, this policy trajectory enabled the government to argue that its position was not only a 'responsible' one but also one that made Australia a key partner of the United States in attempts to prevent the

103 Ibid.
104 Ibid.
105 Ibid., 661.
106 Ibid.
107 Ibid., 664.

further spread of nuclear weapons. And the general proposition that it was through engagement with the global application of nuclear technology that Australia could be most influential in the non-proliferation regime extended beyond the question of uranium exports alone: From 1977, Australia joined the IAEA's Asian Regional Cooperative Agreement (RCA) in Nuclear Science and Technology, initially with an outlay of $100,000 from the foreign aid budget. In his Cabinet submission, Foreign Minister Peacock wrote that

> developing countries … argue that insufficient weight is being given to technical assistance compared with the Agency's [IAEA] safeguards activities. This trend is likely to intensify … Technical assistance will therefore assume increasing importance as a quid pro quo for acceptance by developing countries of tightened safeguards on nuclear materials, equipment and technology.[108]

In their regular decisions to continue Australian membership in the RCA, successive Australian governments would support and re-affirm this argument by the Fraser government – most recently in 2011.[109]

The INFCE and the Future of Australia's Nuclear Ambitions

The subsequent development and implementation of Australia's nuclear safeguards policy would reveal the seriousness with which the Fraser government was committed to achieving a balance between exploiting Australia's uranium resources and upholding its non-proliferation obligations and interests. Its ability to achieve this balance however was seriously tested by the push and pull of technological and political developments at the international level. Of most immediate importance in this respect was the progress of the Carter administration's non-proliferation initiatives. As Wayne Reynolds has argued, the outcomes of the Carter administration's International Nuclear Fuel Cycle Evaluation (INFCE) initiative and the shifting emphases in US nuclear commercial and non-proliferation policy converged to effectively constrain Australia's ability to become anything other than a supplier of natural uranium.[110]

---

108 Cabinet Decision No. 3897_RCA_1977, NAA A12909, 1645.

109 For example, the government's National Interest Analysis, [2011] ATNIF 25, on the *Fifth Agreement to Extend the 1987 Regional Cooperative Agreement for Research, Development and Training related to Nuclear Science and Technology done at Bali on 15 April 2011*, para. 12, states that '[n]on-acceptance of the Fifth Extension Agreement would … diminish Australia's standing in international nuclear arms control fora and our ability to influence international nuclear policy developments in accordance with our national economic and security interests'.

110 Wayne Reynolds, 'The Yellow Cake Road: Malcolm Fraser, the Ranger Enquiry and Australia's Role in the US International Fuel Cycle Project', *Australian Journal of Politics and History* 57, no. 4 (2011): 516–21.

The arguments presented to justify its uranium decision illustrated that the Fraser government believed that as a country with substantial reserves of uranium, 'Australia had acquired a new status and influence in the international community ... that requires it to speak, and commands that it be heard, on issues like proliferation'.[111] The most immediate expression of this belief was the government's eager participation in the INFCE process. The INFCE, established by Carter as a multilateral effort to develop a new consensus on the structure of the international nuclear fuel cycle and appropriate safeguards measures, began its organizational meetings in Washington in October 1977.[112] This initiative was tasked with systematically re-evaluating the nuclear fuel cycle with a particular emphasis on examining: alternative fuel cycles to those based on either the reprocessing of spent nuclear fuel (to obtain plutonium) or uranium enrichment technologies (e.g. gas centrifuge enrichment); methods of spent fuel storage; and improving safeguards for existing technologies.[113] Australia's involvement was significant as it 'sought and secured' the co-chairmanship of 'Working Group 3' of the INFCE, which was concerned, appropriately enough given Australia's interests, with the issue of assurance of long term supply of nuclear fuel.[114] The INFCE, in which some 66 countries participated, was however only a 'technical and analytical' study of the nuclear fuel cycle and potential improvements of nuclear safeguards, and participants would, in the words of the INFCE organizational meeting's final communiqué, 'not be committed to INFCE's results'.[115]

The reasons outlined for Australia's participation in the INFCE, as detailed in Cabinet papers in October 1977, demonstrated both the push and pull of Australia's commercial/economic and non-proliferation interests (particularly at the domestic political level) and the effect of US non-proliferation policy. Thus Cabinet noted on 11 October 1977 that Australia's participation in the INFCE should be governed by three objectives: (i) that Australia should 'participate positively and actively' and 'be seen' to be doing so 'both in Australia and abroad'; (ii) while the areas studied by the INFCE were technically complex and subject to 'national sensitivities and international disagreements' (particularly on the issue of reprocessing), Australia would seek to narrow areas of disagreement so as to 'facilitate the negotiation of bilateral safeguards agreements with uranium

111 Russell B. Trood, 'Australian Uranium Exports: Nuclear Issues and the Policy Process', in Robert Boardman and James F. Keeley, eds, *Nuclear Exports and World Politics: Policy and Regime* (London: Palgrave Macmillan, 1983), 139.

112 Steven J. Warnecke, 'Non-proliferation and the INFCE: An Interim Assessment', *Survival* 21, no. 3 (1979): 117.

113 Philip Gummett, 'From NPT to INFCE: Developments in Thinking about Nuclear Proliferation', *International Affairs* 57, no. 4 (1981): 553.

114 Ibid., 120; Trood, 'Australian Uranium Exports', 138.

115 Final Communiqué of the Organizing Conference of the International Nuclear Fuel Cycle Evaluation (Washington, DC, 21 October 1977).

importing countries which incorporate elements of Australian policy'; and (iii) Australia should encourage INFCE participants to recognize the need for 'a stringent approach towards safeguards of the kind embodied in Australia's safeguards policy'.[116]

Structuring Australia's involvement in this manner would, it was believed, achieve two core goals for the government. First, it would demonstrate to domestic audiences the seriousness with which the government took its non-proliferation obligations 'in the context of the ongoing uranium debate', and second, demonstrate 'that uranium export can positively support our non-proliferation objectives, since it will, in part, strengthen Australia's voice in international nuclear forums'.[117] Cabinet's discussion of the reasons for the government's acceptance of the US invitation to chair Working Group 3 of the INFCE also further underlined the extent to which it had accepted the logic of the Carter administration's approach:

> We based our acceptance on the considerations that by becoming a major uranium exporter, Australia has substantially strengthened its role in international forums working to strengthen the non-proliferation regime and that, by becoming a major exporter, Australia's non-proliferation policy is more readily realised through supply arrangements which might postpone the movement into fast breeder reactors.[118]

However, the findings of the INFCE, when its final report was released in February 1980, ultimately ran counter to many of the preferences of the Carter administration.[119] In particular, the INFCE anticipated the possible commercial use of plutonium within two decades, took the view that states had a 'right' to use plutonium in power generation, and that a state's decision to do so should be based not only on non-proliferation considerations but also economic grounds.[120] In perhaps the starkest rebuke to Carter's agenda, the INFCE 'concluded that fuel cycles cannot be ranked in terms of proliferation risk, based on whether or not they employ reprocessing of spent fuel'.[121] This result clearly undermined the central assumption upon which the Fraser government had constructed its 'uranium decision' and presented Australia with a problem. If the 'plutonium economy' remained both a viable and internationally acceptable option for

116 Cabinet Decision No. 4041, 11 October 1977, NAA A12909, 1712.

117 Ibid.

118 Ibid.

119 For reviews of the INFCE see Warren H. Donnelly, 'INFCE and US Non-Proliferation Policy', *Energy Policy* (December 1980): 335–8; Ted Greenwood and Robert Haffa, 'Supply-Side Non-Proliferation', *Foreign Policy* 42 (Spring 1981): 125–40.

120 Donnelly, 'INFCE and US Non-Proliferation Policy', 337.

121 Greenwood and Haffa, 'Supply-Side Non-Proliferation', 131.

'energy deficient' states, what incentive was there for them to purchase uranium from Australia with 'stringent' safeguards attached? In this respect it is difficult to disagree with Trood's judgment of the bind that the Fraser government thus found itself in when he suggested that there was no evidence 'that the lure of Australian uranium is proving so irresistible' to those states who did not share Australia's non-proliferation commitment 'that they would be willing to compromise options of further nuclear development by agreeing to Australia's conditions of purchase'.[122]

Australia, Uranium Enrichment and Reprocessing

In the midst of the Fraser government's presentation of its nuclear safeguards and uranium export policy, the Cabinet also continued to actively consider options to expand Australia's participation in the nuclear fuel cycle. On 19 July 1977, Cabinet discussed the economic and political feasibility of establishing a uranium enrichment capacity in Australia. Cabinet correctly identified that the major obstacle to an Australian enrichment capability would be access to the required technology, either through independent development or through commercial licensing.[123] Yet the motivation for developing such a capability was not exclusively economic, with Cabinet noting that 'It may be that Australia could best contribute to nuclear non-proliferation by selling uranium in an enriched form rather than in the form of yellowcake.'[124]

Further Cabinet discussion of this issue on the eve of Fraser's 'uranium decision' statement to parliament in August illustrated the linkages the government had made between its economic interests in the exploitation of Australian uranium and its potential to play a positive non-proliferation role. On 1 August 1977 Cabinet agreed that Australia should seriously explore the potential for establishing uranium conversion and enrichment capabilities to 'value add' to nuclear exports and the possibility of inviting an international consortium to establish a 'multinational or multilateral' uranium enrichment facility in Australia.[125] Cabinet saw this as a potential 'major contribution' to global non-proliferation efforts, and the Cabinet decision noted that multinational enrichment plants were 'becoming accepted as one of the most

---

122 Trood, 'Australian Uranium Exports: Nuclear Issues and the Policy Process', 136.

123 Cabinet Decision No. 3364, 19 July 1977, NAA A12909, 1453.

124 Ibid.

125 Cabinet Decision No. 3605, 1 August 1977, NAA A12909, 1539. The document estimated that the establishment of a conversion plant would increase export sales values from $360 million to 'about $380 million', while the establishment of an enrichment capacity would raise the value of uranium exports from $360 million to 'about $700 million'.

promising routes to promote a diverse nuclear fuel supply industry while minimising proliferation hazards'.[126]

Crucially, and despite the clear thrust of the Carter administration's non-proliferation policy, Cabinet continued to believe that 'President Carter's policies do not rule out the possibility of establishing enrichment capacity in Australia.'[127] This demonstrated the continued tenacity of what might be termed the national development perspective on Australian nuclear policy that sought to fully exploit the economic and technological potential of the country's uranium resources. Australia's desire to engage in the full range of nuclear fuel cycle activities, however, very clearly ran counter to the trajectory of the Carter administration's approach to non-proliferation and the nuclear fuel cycle: While US Secretary of State Warren Christopher stated that access to Australia's uranium 'could be critical in terms of the global supply picture and our efforts to delay premature entry into the plutonium economy', the United States was not prepared to export 'sensitive technology, including that needed for enrichment and reprocessing'.[128]

With unpromising signals from the Carter administration, the Fraser government conducted discussions with the French and Japanese governments and representatives of the European consortium URENCO between February and April 1979 regarding the feasibility of establishing a commercial enrichment industry in Australia.[129] In parallel, the government also approached four Australian companies – BHP, CSR, WMC and Peko-Wallsend – that 'had previously expressed an interest in the development of a commercial uranium enrichment industry'.[130] Discussions between government officials and representatives of the four Australian companies, which took place between May and August 1979, ultimately led to the formation of the Uranium Enrichment Group of Australia (UEGA) to study the feasibility of a commercial enrichment industry. On 2 January 1980 the Deputy Prime Minister and Minister for Trade and Natural Resources, Doug Anthony, and the Minister for National Development and Energy, John Carrick, released a joint public statement on the progress of the government's study of the feasibility of commercial enrichment. The statement noted the government's discussions with the United States, France, Japan and URENCO regarding the 'possibility of enrichment technology being made available to Australia' were proceeding 'satisfactorily'. It also positively noted the formation of UEGA and affirmed the government's commitment that Australian industry would play 'a major

126 Ibid.

127 Ibid.

128 Memorandum for the President from Warren Christopher, 'Visit of Australian Minister Fraser', 16 June, cited in Reynolds, 'The Yellow Cake Road', 515.

129 See Cabinet Decision No. 8946, 12 June 1979, NAA A12909, 3204.

130 Cabinet Decision No. 10717, 18 February 1980, NAA A12909, 3797.

role … in studying, and if the venture proves feasible, developing an enrichment industry in Australia' as the 'Government would be concerned to see that the arrangements resulted in the maximum economic benefits of that industry being retained in Australia for the benefit of Australians'.[131] The statement concluded with what had become the Fraser government's boilerplate on uranium-related issues, arguing that while Australia's natural endowment of uranium made it 'well placed to proceed with uranium enrichment' the government would give 'detailed consideration' to a variety of potentially problematic issues such as 'the most appropriate technology to be employed', 'environmental factors' and 'questions of nuclear non-proliferation and safeguards'.[132]

In the wake of the release of the final INFCE report in February 1980, Australia's quest for a uranium enrichment capability was given further impetus. The INFCE's conclusions that fuel cycles could not be ranked in terms of proliferation risk and that, as plutonium was an inevitable by-product of the operation of nuclear reactors, the core problem was not how to avoid that production but rather how to manage it, represented a clear rebuff to the Carter administration.[133] Additionally, the INFCE also projected that the world's uranium enrichment capacity was sufficient to meet world demand 'up until about 1990'. According to a Cabinet discussion paper tabled by Minister for Foreign Affairs, Andrew Peacock, and Deputy Prime Minister and Minister for Trade and Resources, Doug Anthony, in March 1980, these conclusions helped 'support Australia's case for considering the establishment of a commercial uranium enrichment industry to serve the world market in the 1990s'.[134]

Such optimism was given further strength by two developments. First, the relative reticence of the United States vis-à-vis provision of uranium enrichment technology to Australia promised to be overturned with the defeat of President Carter in the November 1980 presidential election at the hands of Ronald Reagan. In particular, the incoming administration was especially critical of what a number of observers suggested was the Carter administration's 'universalism' in the application of the US 'technology denial' approach to nuclear proliferation.[135] The likelihood of US policy change on this front was in fact flagged with Australian officials. For example, in a letter to UEGA Chairman Dr R.G. Ward on 4 February 1981, the Acting First Assistant Secretary of the

131 Ibid., 'Attachment A: Uranium Enrichment Study', 2 January 1980.

132 Ibid.

133 Haffa and Greenwood, 'Supply-Side Nonproliferation'.

134 Cabinet Submission No. 3835, 11 March 1980, NAA A12909, 3835.

135 See Gerard Smith and George Rathjens, 'Reassessing Nuclear Nonproliferation Policy', *Foreign Affairs* 59, no. 4 (1981): 880–82; Frank B. Cross and Cyril Smith, 'The Reagan Administration's Nonproliferation Nonpolicy', *Catholic University Law Review* 33 (1983/84): 640; Warren H. Donnelly, 'Nonproliferation Policy of the United States in the 1980s', *SAIS Review* 7, no. 2 (1987): 167–71.

Department of National Development and Energy outlined a meeting between Unites States and Australian officials at the Australian Embassy in Washington regarding the issue of US/Australian collaboration on uranium enrichment. The letter stated that 'the policy considerations raised by the matter would be dealt with expeditiously by the new Administration whose approach to the range of nuclear matters would be quite different from that adopted by the Carter Administration'.[136]

President Reagan's statement of his administration's non-proliferation policy on 16 July 1981 confirmed that the United States would indeed be changing tack. Encapsulating his administration's core disagreement with the previous administration on non-proliferation, Reagan argued that:

> We must reestablish this Nation as a predictable and reliable partner for peaceful nuclear cooperation under adequate safeguards. This is essential to our nonproliferation goals. If we are not such a partner, other countries will tend to go their own ways, and our influence will diminish. This would reduce our effectiveness in gaining the support we need to deal with proliferation problems.[137]

Ultimately the administration's position would be driven by three core interests or principles flagged in this statement: re-establish the United States as a 'reliable' supplier of nuclear material and technology; a discriminatory approach to nuclear export policy that considered a potential recipient country's technological status and proliferation risk; and an emphasis on security measures to reduce proliferation incentives.[138] Significantly, Australia, given its long-standing alliance with the United States and its demonstrated commitment to non-proliferation and nuclear safeguards, under these criteria would be looked upon favourably by the administration as a potential recipient of nuclear technology.

In December 1980 the Chairman of UEGA, R.G. Ward, provided Prime Minister Fraser with recommendations from the group's interim feasibility study. The most significant were the conclusion that 'the establishment of a commercial enrichment industry in Australia could be both feasible and sufficiently profitable to be attractive to private industry'; that 'limited markets for enrichment services should be available from about 1990'; and that centrifuge technology 'developed independently by Japan, the United States

---

136 'Letter to Dr R.G. Ward from the Department of National Development and Energy', 4 February 1981, NAA A1838, 720/4/9 Part 39.

137 Ronald Reagan, 'Statement on United States Nuclear Nonproliferation Policy', 16 July 1981, *Public Papers of President Ronald Reagan*, available at http://www.reagan.utexas.edu/archives/speeches/1981/71681a.htm [accessed 2 July 2014].

138 See Ross and Smith, 'The Reagan Administration's Nonproliferation Nonpolicy', 641; Donnelly, 'Nonproliferation Policy of the United States', 170.

and URENCO/CENTEC is the most attractive economically'.[139] In April 1981 UEGA reported to the government that it would be in a position to make a recommendation by the close of the year regarding a preferred technology on which to base 'a full feasibility study'.[140]

The report, delivered in September 1982, noted that the new president of the United States had 'authorized US companies to provide additional and updated confidential commercial information on US centrifuge technology to UEGA', but was nonetheless sceptical that US technology would be a commercially viable option given the 'limited prospects for access to US markets, possible restrictions on Australian manufacture of some components and possible complications in obtaining Congressional approval for technology transfer'.[141] Given these considerations, UEGA recommended that URENCO should be the preferred option as their centrifuge technology was economically superior and had a proven technical and commercial capability.[142] The Fraser government accepted this recommendation and agreed with the UEGA's position that a commercial agreement between UEGA and URENCO be finalized 'before the commencement of a detailed engineering and feasibility study'.[143] The electoral victory in March 1983 of the ALP under Bob Hawke ultimately halted this process. In October 1983 the new government decided to inform UEGA that 'at this time the Government does not wish to further facilitate the UEGA study on uranium enrichment or be otherwise involved in the study' on the basis of the government's commitment to the ALP platform that 'prohibits the enrichment of uranium in Australia'.[144]

Conclusion

Prime Minister Fraser saw Soviet strategic belligerence, abetted by the West's misplaced faith in détente, as a central global strategic challenge. Particularly worrisome from the government's perspective was the spectre of the Soviets achieving strategic superiority in the nuclear arms race, as this could embolden Moscow's expansionist tendencies with deleterious consequences for international and regional security. However, while the central nuclear balance was of importance for Australia, its ultimate effect on the country's security

139 Cabinet Decision No. 13880, 17 February 1981, NAA A12909, 4541, Attachment A.

140 Cabinet Decision No. 19070, 5 October 1982, NAA A12909, 5784.

141 Ibid., Attachment B: Investigation of the Establishment of an Uranium Enrichment Industry Interim Report on Technology Choice and Plant Siting and Other Matters, September 1982.

142 Ibid.

143 Ibid.

144 Cabinet Decision No. 2355, 31 October 1983, NAA A13977, 246.

was indirect, through its effect on regional conflicts. Here, global strategic concerns intersected with Australia's interest in curbing nuclear proliferation, which had the potential of making regional conflicts far more dangerous and difficult to manage.

Australia's commitment to the NPT as an NNWS was thus in large measure driven by its desire to see 'that as many countries as possible ratify the Non-Proliferation Treaty', as this would assist in limiting nuclear proliferation in its immediate region.[145] Thus Australian commitment to the NPT, in contrast to other US allies particularly in Europe, was made not on the basis that it would help stabilize the superpower confrontation, but because it would enhance Australia's security in its own strategic environment. Balancing Australia's interest in strengthening the NPT among the NAM, with its interest in US nuclear strength vis-à-vis the Soviet Union, meant that Australia did not always see eye-to-eye with the other members of the 'Western group' on issues such as the CTB negotiations – in particular, until the end of the 1980 NPT RevCon.

Overall, the Fraser years constituted a watershed moment in the history of Australia's engagement with the nuclear age. Australia had by the end of the Whitlam government ratified the NPT as an NNWS. Its commitment to this cornerstone of the emerging non-proliferation regime did not however diminish national policy debates about the utility of emerging nuclear arms control and non-proliferation architecture on the one hand, and the implications of this for the exploitation of Australia's uranium resources on the other. The Fraser government's ability to strike a 'grand bargain' between Australia's interest to limit the further spread of nuclear weapons and its desire to exploit the economic potential of Australia's uranium resources was complicated by domestic political considerations, but also significant changes in the strategic, normative and economic domains of nuclear policy at the global level. Its nuclear policy was an attempt to balance Australia's interests across these domains, but the conditions that made the 'grand bargain' possible were ultimately of a transitory nature.

Australian uranium export policy, it was argued, helped strengthen the non-proliferation regime by providing a tangible benefit – i.e. access to Australian uranium – to states that abided by the treaty. The four 'cornerstones' of the Fraser government's uranium export policy – strict adherence to the NPT, IAEA safeguards, bilateral agreements between Australia and importing countries and Australian participation in multilateral efforts to strengthen safeguards and the non-proliferation regime – operationalized these concerns and established a position that Australia would couple a commitment to the NPT as the centrepiece of the international non-proliferation regime, with an

---

145 Defence Committee, *Strategic Basis of Australian Defence Policy*, 3 October 1975, para. 171.

increasingly activist role in multilateral non-proliferation regimes. The Fraser government's basic position that Australian uranium gave it the capacity for significant influence to strengthen the non-proliferation regime – and that this was ultimately in Australia's own strategic interest – was one that would become the bipartisan orthodoxy in later years.

Chapter 4
Nuclear Challenges between Washington and Wellington: The Hawke Government (1983–91)

The Labor government under Robert (Bob) Hawke came to power in international circumstances that were still dominated, until the late 1980s, by intense Cold War competition. NPT membership increased during the 1980s by 30 countries to 140 in 1989, but most of the increase was due to newly independent, small Caribbean and Pacific countries joining the treaty. Universal membership remained a distant objective, and the question of nuclear exports to non-member states continued to be an internationally vexed issue. Nonetheless, under the new government Australia began to pursue a more ambitious global arms control agenda. Australia's conception of its nuclear role arguably shifted towards being a 'middle power white knight', a state committed to pursuing practical outcomes on arms control (including non-proliferation). The appointment of career diplomat Richard Butler as Australia's inaugural Ambassador for Disarmament exemplified this endeavour to craft a more activist and internationalist profile for Australia on nuclear issues – overseas as well as at home.

At the same time, however, the new government also was conscious that it would have to develop a defence policy seen as viable by the general population if it was to demonstrate its governing credentials. This required the Hawke Cabinet to reconcile some acute challenges in the nuclear domain, where the government was caught between the left wing of the Labor Party and a rapidly growing grassroots anti-nuclear movement on the one hand, and the expectation of the Reagan administration that Australia would support Washington on nuclear issues on the other hand. This expectation became more pressing following the Lange government's decision to prohibit the entry of nuclear-armed vessels into New Zealand ports, and Washington's decision to suspend New Zealand from the ANZUS Treaty.

Hawke personified the tension between competing domestic and international pressures on Australian nuclear policy. On the one hand, the Prime Minister was by instinct pro-alliance in his outlook, and his credentials as a supporter of the United States were well established. On the other hand, Hawke was a consummate grassroots politician who sensed the shifting winds of public opinion at home, which indicated increasing domestic support for nuclear arms control and disarmament, even if it meant offending Australia's most important ally, the United States. But although these conflicting demands required at times difficult diplomatic and political trade-offs, the government confronted them in strategic circumstances in which the salience of nuclear weapons for Australia's immediate security already began to fade.

Australia and the Superpower 'Arms Race'

Throughout the Cold War, Australia had always been geographically removed from the areas of direct superpower confrontation in Europe and North Asia. Concerns in Australia about the nuclear balance during the 1970s arose from the perception that the Soviet Union held the initiative, both in strategic nuclear force expansion and in extending its influence in the third world, and that these two developments were ultimately linked. Restoring US nuclear capabilities to (perceived) parity was seen by many as the first step necessary to stabilize nuclear competition in the 1970s, and arms control agreements, including the CTB, were sometimes in conflict with this goal.

When the Hawke government assumed office in March 1983, however, the situation was quite different. The Reagan administration was engaged in a large effort to make the Afghanistan War into the Soviet Union's Vietnam; opponents of Communist forces and governments in Africa and Central America received significant US military assistance; and a US intervention would even directly depose the Cuban-aligned government of Grenada in November of the same year. At the nuclear level, the Reagan administration's defence build-up included support for major programs such as the B-1 bomber, MX and Midgetman missiles, as well as the Strategic Defense Initiative. Even if the administration went about nuclear force modernization in a somewhat haphazard manner, there was little doubt that it did reverse perceptions of a decline in US capabilities and will to prosecute the Cold War. This was reinforced when NATO governments in November 1983 faced down domestic opposition and Soviet threats, and began to deploy new nuclear Pershing II and ground launched cruise missiles to several countries in Western Europe, consistent with their double-track decision of 1979.

Reduced concern that the nuclear competition would give openings for Soviet (conventional) adventurism in the third world, which could then cause direct threats to Australia, was evident from the Australian Defence Committee's 1983 *Strategic Basis* paper. Although it judged that the Soviet Union had strategic nuclear superiority over the United States, and noted that it was in Australia's interest for the 'correlation of world forces' not to shift away from the United States and its allies, concerns about new areas of confrontation with the Soviet Union focused on the Middle East and Pakistan.[1] In contrast, 'Australia's own

1 Defence Committee, *Strategic Basis of Australian Defence Policy* (Canberra: Australian Department of Defence, 15 August 1983), paras 5–6, 16–19, NAA A13977, 392.

neighbouring regions are not at this time – and seem unlikely to become – theatres of significant confrontation between the super-powers.'[2] The Soviet naval presence at Cam Ranh Bay was seen in the context of its client relationship with Vietnam, and the harbour's role as a staging post for deployments to the Indian Ocean, rather than as a tool to exert pressure on other Southeast Asian countries.[3] And in the 1987 Defence White Paper, the government itself would likewise not draw any link between the Soviet presence in the region and the central nuclear balance.[4]

By 1983 the international context of the Cold War was thus such that international concerns about the nuclear competition mostly focused on reducing political tensions and the nuclear 'arms race', rather than on Western weakness in nuclear forces. The first two years of the Hawke government, until Mikhail Gorbachev became general-secretary of the Communist Party of the Soviet Union in November 1985, were particularly apt to reinforce fears of nuclear war between the superpowers. In that sense, Australia's domestic debate, and that inside the left wing of the political spectrum in particular, mirrored concerns that were also shared in the Northern hemisphere. Rather than Soviet aggressiveness, which nuclear weapons might help reduce through deterrence, the practice of nuclear deterrence itself was seen by many as the main source of danger. In this context, the view that the only purpose of nuclear weapons was to counter other nuclear arsenals – a perspective evident in the 1986 Joint Parliamentary Committee's report on *Disarmament and Arms Control in the Nuclear Age*, and also one shared by Minister for Foreign Affairs Gareth Evans, became widespread even within official circles in Australia.[5]

Overall, however, the Hawke government's foreign policy would be characterized by continuity with that of its predecessor, as Hawke himself acknowledged in a public speech in Washington in 1983. Australia would be 'neutral neither in thought, nor in action', but its foreign policy would be 'independent and self-respecting'. Crucially, Hawke claimed that hosting the joint facilities gave Australia 'an entitlement to be heard' about the strategic balance.[6] A hallmark of the Hawke government's approach would be to leverage strong support for the United States for its arms control agenda, and vice versa, in Washington as well as in the Labor caucus and party.

During a visit to Washington, Hawke had privately confirmed Australian support to the MX test in 1984. The Security Committee of Cabinet agreed in

2 Ibid., para. 7.

3 Ibid., para. 31.

4 Australian Department of Defence, *The Defence of Australia* (Canberra: Australian Government Publishing Service, March 1987), 14–15.

5 See Lyon, 'Australia: Back to the Future?', 438–9.

6 Rt. Hon. Robert Hawke, 'Speech by the Australian Prime Minister to the Washington Press Club' (Washington Press Club, Washington, DC, 15 June 1983).

January 1985, after the test had been moved out of the Australian economic zone into international waters, and decided to refer to the Fraser government's commitment should the test become public knowledge.[7] When the decision was leaked to the press, Hawke, Kim Beazley and other Cabinet ministers publicly defended it for several days against a revolt in caucus and the wider party, knowing that the United States at the time was acutely concerned about anti-nuclear sentiment amongst its allies in general, and New Zealand's refusal to allow ship visits in particular.[8] In the end, personal discussions between Prime Minister Hawke and US Secretary of State George P. Shultz led to changes to the test program that would eliminate the need for Australian support.[9] The circumstances of the MX decision temporarily cost the government political support at home, but it was ultimately without any practical impact on the US program. Australian critics on the left had won a short-term victory, but government ministers had also demonstrated to the United States where their policy preferences lay.

This early crisis thus made it easier for the Hawke government to manage tensions about other nuclear policy issues in following years. For example, there was considerable opposition in the Labor party to the Reagan administration's Strategic Defense Initiative, which was seen to undermine the condition of 'mutual assured destruction' with the Soviet Union. The Hawke government was thus placed in an awkward situation when it became public that US Secretary of Defense Caspar Weinberger had written to Australia (and other US allies) in March 1985 inviting participation in research towards the program. In the lead-up to the ministerial AUSMIN meetings with the United States to be held in August, Defence Minister Kim Beazley publicly declined this invitation during Question Time in Parliament in April 1986.[10] There was, however, little specific to Australia in this response as many European allies were equally sceptical towards the initiative, which was subject to lively debate in the United States as well as in NATO.[11] Australia was hardly breaking ranks on this issue.

The MX debate also gave additional credence to the Hawke government's policy on the Joint Facilities. Foreign Minister Hayden had laid much of the

7 Cabinet Decision No. 4613, 29 January 1985, NAA A13979, 4613.

8 A link that was also made at the time. See Terry Atlas and James O'Shea, 'The U.S. to Test MX without Australian Aid', *Chicago Tribune*, 7 February 1985, available at http://search.proquest.com/docview/290820828?accountid=8330 [accessed 20 July 2014].

9 Blanche D'Apulget, *Hawke: The Prime Minister* (Melbourne: Melbourne University Press 2011), 99–110.

10 *C.P.D.* (Representatives), 8 April 1986, 1795.

11 See for example Andrew White, 'European Perspectives on the Strategic Defense Initiative', *Millennium-Journal of International Studies* 15, no. 2 (1986), 211–22; Nanette C. Brown, *The Strategic Defense Initiative and European Security: A Conference Report* (Santa Monica, CA: RAND, 1986).

domestic groundwork by justifying their continued existence in public based on their role in arms control and deterrence stability, especially in a paper, published by the Department of Foreign Affairs and Trade, entitled *Uranium, the Joint Facilities, Disarmament and Peace*.[12] After Kim Beazley became Minister for Defence in 1984, he led many of the subsequent debates.[13] In parallel, the Hawke government conducted negotiations with the United States, in which it argued for a greater say and involvement in the operation of the facilities. This was based on general arguments about sovereignty, but it also stemmed more specifically from a concern that the intelligence gathered by each installation could be used to support US nuclear first-strike options, rather than bolstering deterrence (as it was understood in Canberra) with the Soviet Union.[14]

Under new arrangements that Hawke explained in Parliament in 1988, Australian staff would assume increased responsibility for the management of Nurrungar and Pine Gap, including as deputy commanders. In addition, Australians would undertake an increasing share of the actual operational work, thereby genuinely gaining 'full knowledge and concurrence' of their daily use.[15] Australia continued to lend its support to US nuclear deterrence, but the Hawke government managed to gain recognition from the United States that there were also specifically Australian interests involved in doing so. Domestically, this ultimately reconciled the Labor Party with the existence of the Joint Facilities in Australia.

The Fading of the Domestic Nuclear Option

Ratification of the NPT by Australia in 1972 did not mean that the option of nuclear weapons acquisition by Australia disappeared as a policy consideration. There was a possibility that signatories could decide to let the treaty lapse after 25 years, and any party could leave the treaty under the 'supreme interests' clause.[16] In current and foreseeable circumstances Australia was better served by membership in the NPT, but that judgment was something that could change: Even before the country signed the treaty, the Defence Committee had argued in the 1968 *Strategic Basis* that:

> should a serious breakdown in the international order appear likely to develop, Australia might wish to reconsider the possibility of a requirement for a

12 Australian Department of Foreign Affairs, *Uranium, the Joint Facilities, Disarmament and Peace* (Canberra: Australian Government Publishing Service, 1984).

13 D'Apulget, *Hawke: The Prime Minister*, 103–4.

14 Interview with Kim Beazley, Washington, DC, 31 March 2011.

15 See *C.P.D.* (Representatives), 22 November 1988.

16 Thomas Graham, 'The Duration of the Nuclear Non-Proliferation Treaty: Sudden Death or New Lease on Life?', *Virginia Journal of International Law* 29, no. 3 (1989): 661–77.

> nuclear capacity. It is important, therefore, that Australia maintain its freedom to reduce the lead time for the development of such a capacity from the present period of from seven to ten years.[17]

Throughout the 1970s, similar comments were included in subsequent *Strategic Basis* papers.[18] None of the Cabinet decisions on the papers or the Defence Ministers' submissions recorded objections or qualifications to these considerations. Australian lead time, the financial and industry requirements for a nuclear weapons program, and possible nuclear force structure options were analyzed in policy papers, for example by academics working at the Australian National University's Strategic and Defence Studies Centre in 1972, 1974 and 1982.[19] In 1976, the Fraser Cabinet's only major change of substance to the draft of the 1976 *Australian Defence* White Paper was to delete a sentence that read

> We see no requirement in our present and prospective strategic circumstances to acquire nuclear weapons. Such a move would also cause concern among many nations of importance to us; it would impose a heavy and at present unnecessary burden on our defence effort.[20]

In its 1981 report on *Threats to Australia's Security: Their nature and probability*, the parliamentary Joint Committee on Foreign Affairs and Defence (JCFD) was less circumspect: in a section on 'issues that need to be addressed' in relation to Australia's security from global conflict, a long list of questions included the following:

---

17 Ibid., 387.

18 See Defence Committee, *Strategic Basis of Australian Defence Policy* (Canberra: Australian Department of Defence, 5 March 1971), para. 192; Defence Committee, *Strategic Basis of Australian Defence Policy*, 1 June 1973 (Canberra: Australian Department of Defence), para. 40; Defence Committee, *Strategic Basis of Australian Defence Policy*, 3 October 1975, paras 39, 264; Defence Committee, *Australian Strategic Analysis and Defence Policy Objectives*, paras 96, 382. Of these papers, the Defence Committee was most circumspect about the possibility of a future nuclear weapons program in the 1973 *Strategic Basis* submitted to the Whitlam Government.

19 Ian Bellany, *Australia in the Nuclear Age: National Defence and National Development* (Sydney: Sydney University Press, 1972); Robert O'Neill, *Internal Problems Associated with any Acquisition of Nuclear Weapons by Australia* (Canberra: Australian National University, Strategic and Defence Studies Centre, 1974); Desmond Ball, 'Australia and Nuclear Policy', in Desmond Ball, ed., *Strategy & Defence* (Sydney: George Allen & Unwin, 1982), 320–42. See also Stephan Frühling, 'Never Say Never: Considerations about the Possibilities of Australia Acquiring Nuclear Weapons', *Asian Security* 6, no. 2 (2010), 146–69.

20 Cabinet Decision No. 1656 (FAD), 18 October 1976, NAA A12909, 703.

> Is there any likelihood of a situation where Australia would have to acquire nuclear weapons? The possible counter-productive consequences of any steps towards this, in stimulating a regional arms race, would need to be carefully considered.[21]

When the Hawke government considered its first (and last) *Strategic Basis* paper in 1983, the Defence Committee again recommended that:

> Developments relating to nuclear capability in countries within Australia's neighbourhood should be monitored in order to ensure that the lead time for Australia could be matched with developments in other countries should Government so decide.[22]

The Defence Minister, Gordon Scholes, noted in his submission to Cabinet that the Defence Committee 'sees no major implications in [multi-national issues] for our defence planning (although we must ensure that we can assess and keep abreast of regional nuclear developments)'.[23] Thirteen years later, Scholes mentioned in an interview that '[t]he proposition was that we should maintain our research or knowledge to a level which would enable us to move if, say, Indonesia moved in that direction', going on to state that '[t]here was a decision by Hawke, [Foreign Minister] Hayden and me along those lines and I was to prepare a letter which was to be sent to Hawke if it became necessary'.[24] Hayden himself stated in his autobiography that in November 1984 he proposed 'to a very small group of Ministers, including the Prime Minister', that the government institute research so that Australia could reach 'the threshold of being able to assemble nuclear weaponry … in the shortest possible time'.[25] And yet, the discussions by Scholes and Hayden are probably the last time that an Australian domestic nuclear weapons option, however remote, was the subject of serious discussions by government. As Hayden himself noted, he was under the 'misapprehension' that his 'views had been accepted', as 'nothing was subsequently done about the proposal'.[26]

Several factors contributed to the fading of the domestic nuclear option from Australian defence policy discussion under the Hawke government.

---

21 Commonwealth of Australia, 'Threats to Australia: Their nature and probability', *Joint Committee on Foreign Affairs and Defence* (Canberra: Australian Government Publishing Service, 1981), 29.

22 Defence Committee, *Strategic Basis of Australian Defence Policy*, 15 August 1983, NAA A13977, 392.

23 Cabinet Decision No. 2165/2131, 27 September 1983, NAA A13977, 392.

24 Gordon Scholes, quoted in M. Steketee, 'Hawke Advisers Drew N-Arms Plan', *The Weekend Australian*, 6–7 April 1996.

25 Bill Hayden, *Hayden: An Autobiography* (Sydney: Harper Collins Publishers, 1996), 422–3.

26 Ibid., 423.

One of these was bureaucratic: The *Strategic Basis* papers tended to be wide-ranging in their scope, classified and only intended to inform subsequent policy decisions. Hence, the Defence Committee also discussed remote contingencies, such as a possible decision to acquire nuclear weapons that did not require any practical action, or even a government decision. Scholes himself, for example, made clear that the subject had been brought up with the Defence Committee, and that he 'would have opposed it in any circumstances'.[27] The Hawke government soon experienced one of the disadvantages of the *Strategic Basis* approach to strategic guidance, when the 1983 version, including comments relating to Australian nuclear weapons lead time, was leaked to the press.[28]

The 1983 *Strategic Basis* would however be the last such document. From 1987, defence strategic guidance in Australia was instead contained in Defence White Papers, which were directly statements of government policy and linked the analysis of strategic circumstances to the government's priorities for force structure and defence expenditure. But whereas governments tended to accept the musings of their advisors on far remote and unlikely contingencies in the *Strategic Basis* papers, they generally have little reason to devote much attention to them in the *White Papers*, which they approach as statements of their own policy and priorities.

Another contributing factor was that confidence in the basic security of Australia also increased during the 1980s. The possible need to defend Australia without direct allied support had been the cause of profound strategic anxiety in some circles after the Vietnam War, especially among those who had lived through the fall of Singapore, Japanese attacks on Darwin, Communist expansion throughout mainland Asia, the Indonesian Crisis and Confrontation. As the Australian defence organization and defence services began to pay increased attention to the defence of the continent itself, however, the positive aspects of Australia's remoteness from major powers in the northern hemisphere became increasingly apparent.[29] Hence, even the 1981 JCFD report that raised the question of an Australian nuclear force, remarked that its 'conclusions as to the nature and probability of threats to Australia's security are cautiously reassuring'.[30] Although the 1983 *Strategic Basis* raised the nuclear lead time issue, its

27 Gordon Scholes, quoted in Steketee, 'Hawke Advisers Drew N-Arms Plan'.

28 'The Strategic Basis Papers', *The National Times*, 30 March–5 April 1984; Brian Toohey, *Book of Leaks* (Sydney: Angus & Robertson, 1987), 241–68.

29 Alan Thompson, *Defence Down Under: Evolution and Revolution 1971–88*, Working Paper 40 (London: Sir Robert Menzies Centre for Australian Studies, University of London, 1988).

30 Commonwealth of Australia, 'Threats to Australia: Their Nature and Probability', *Joint Committee on Foreign Affairs and Defence* (Canberra: Australian Government Publishing Service, 1981), 54.

overall conclusions recommended paying closer attention to practical questions of the defence of Australia and Papua New Guinea against regional threats.[31] This was indeed to be the focus of the 1986 Dibb Review of Australia's defence capabilities,[32] and of the 1987 *Defence of Australia* White Paper, which further de-emphasized preparations for high intensity conflict. Former Labor Foreign Minister Gareth Evans even went so far as to call the White Paper a 'conceptual watershed in Australian foreign policy', whose 'intellectual breakthrough was to establish that self-reliance was possible in most of the circumstances likely to confront Australia, short of major war'. As major war was seen as remote and unlikely, this 'liberated Australian foreign policy' from its focus on great power protection.[33] A logical consequence was that there would then also be less need to contemplate replacing that protection with nuclear weapons of its own.

The fading of the domestic nuclear option as a policy consideration of any practical relevance meant that remaining vestiges of the old policy also progressively disappeared. Australian F-111C aircraft, for example, still had the nuclear weapons wiring and release panels installed that were part of the US F-111A version they were based on (and the original Australian requirement for the aircraft in the early 1960s had called for a nuclear-capable aircraft). When the RAAF focused more on the maritime strike role in the defence of Australia, however, the nuclear panel and wiring was removed in the mid-1980s to make way for the instrumentation of the *Harpoon* anti-shipping missile.[34] (A decision that, according to then Defence Minister Kim Beazley, Foreign Minister Hayden did not applaud.)[35] And when the government revised the legislative basis for nuclear policy and research in Australia, as discussed below, any reference to defence purposes was expunged from domestic legislation.

Nuclear Arms Control and the Rarotonga Treaty[36]

A key plank of the Hawke government's attempt to demonstrate its domestic and international credentials was an intensification of Australian efforts in

---

31 Defence Committee, *Strategic Basis of Australian Defence Policy*, 15 August 1983, paras 35, 36.

32 The review itself simply stated that 'There is no foreseeable requirement for Australia to consider acquiring nuclear weapons.' Paul Dibb, *Review of Australia's Defence Capabilities* (Canberra: Australian Government Publishing Service, 1986), 31.

33 Gareth Evans and Bruce Grant, *Australia's Foreign Relations in the World of the 1990s*, 2nd edition (Melbourne: Melbourne University Press, 1995), 29–30.

34 Mark Lax, *From Controversy to Cutting Edge: A History of the F-111 in Australian Service* (Canberra: Air Power Development Centre, 2010), 26, 168, 197.

35 Kim Beazley, pers. comm. to Andrew O'Neil.

36 This section draws on Andrew O'Neil, 'The South Pacific Nuclear Free Zone Treaty: A Reinterpretation', *Australian Journal of Political Science* 39, no. 3 (2004): 567–83.

the area of multilateral arms control. As outlined in the previous chapter, the Fraser government had been committed to pursuing a CTB treaty in the UN Conference on Disarmament. It however significantly reduced the ardour with which it pursued this objective after the 1980 NPT Review Conference, and ultimately had little to show for its efforts when it lost office in March 1983. Like its predecessor, the Hawke government portrayed the conclusion of a CTB as central to international efforts to constrain nuclear proliferation, with Hawke himself noting in 1984 that a treaty 'would help limit the development of new nuclear weapons and the improvement of existing nuclear weapons'.[37] And yet, it too was faced with the same unfavourable international climate in which to push Australia's case for a treaty.

Following the collapse of the trilateral test ban negotiations in 1980, there seemed to be little prospect of a renewal of negotiations between the nuclear powers. By 1982, it had already become clear that the United States remained opposed to any negotiations on the issue. Alleging that Moscow had failed to comply with the 1974 Threshold Test Ban Treaty and 1976 Peaceful Nuclear Explosions Treaty, the United States expressed fears that the Soviet Union would not fully abide by any CTB either.[38] Privately, the Reagan administration was also apprehensive that American engagement in any negotiating process would stall its ambitious nuclear force modernization program.[39]

In 1983, the Hawke government abstained on a resolution in the UN General Assembly calling for a freeze on the testing, manufacture and deployment of nuclear weapons. The following year, however, Australia was one of only three Western aligned countries to vote in favour of the nuclear freeze resolution.[40] This usefully served to publicly highlight the policy differences between the Hawke government and the United States on the CTB issue, and the formation of the Nuclear Disarmament Party and its strong showing in polling in 1984 was essential background to explaining Australia's changed vote. In practice, however, Australia helped to formulate a 'compromise' mandate for the Western Group in the CD that proposed discussions on verification but did not endorse a resumption of negotiations on a treaty text. This was very close to the American position and only served to further delay the process of negotiations on a CTB treaty. In seeking to address American sensitivities, the Hawke

37 *C.P.D.* (Representatives), 6 June 1984, 2986.

38 R.J. Smith, 'Administration at Odds over Soviet Cheating', *Science* 228, no. 4700 (10 May 1985): 695–6.

39 As Simpson has revealed, American opposition to progress in CTB negotiations was closely linked to the Reagan administration's decision in late 1982 to authorize a series of underground nuclear tests. See Christopher Simpson, *National Security Directives of the Reagan and Bush Administrations: The Declassified History of US Political and Military Policy, 1981–1991* (Boulder, CO: Westview Press, 1995), 73–4.

40 Daniel Mulhall, 'Australia and Disarmament Diplomacy, 1983–1985: Rhetoric or Achievement?', *Australian Outlook* 40, no. 1 (1986), 34–5.

government thus continued to advocate a position that proved unacceptable to non-Western Group members of the CD. Rhetoric aside, its policy thus remained similar in substance to that of its predecessor.

Australia's difficulty to achieve progress on test ban issues was however only part of the nuclear arms control record of the Hawke government, which achieved greater success with the conclusion of the South Pacific Nuclear Free Zone (SPNFZ) Treaty in 1986.[41] The new Labor government had entered office in 1983 committed to negotiating a nuclear free zone in the South Pacific. This commitment had its origins in the Whitlam government's earlier push for such a zone, which was closely associated with its decision in 1973 to prosecute (with New Zealand) an International Court of Justice (ICJ) case against French nuclear testing in the South Pacific.[42] Indeed, Labor's commitment to a SPNFZ could be traced back to the 1960s when in opposition it supported the creation of a nuclear free zone in the southern hemisphere.[43]

The Treaty was endorsed by consensus at the 1985 South Pacific Forum in Rarotonga and opened for signature.[44] From the Hawke government's perspective, the conclusion of the Rarotonga Treaty (as it became known) was the successful culmination of more than two years of Australian diplomatic activity. The Treaty embodied all of the basic principles formulated by Australia on issues of concern to the Hawke government, such as maintaining essential freedom of transit for American nuclear platforms through Australian ports and airfields. As Hawke noted at a media conference during the course of the Rarotonga Forum meeting, Australia's alliance obligations were 'perfectly consistent with the provisions of the Treaty'.[45]

The Treaty also served another important purpose for Australia's nuclear policy. Given the time and energy invested by the Hawke government in fostering the SPNFZ negotiating process,[46] and its central role in determining

41 For a detailed analysis of the SPNFZ negotiations that challenges the claim that Australia was simply motivated by protecting the transit of US nuclear platforms across the South Pacific, see Andrew O'Neil, 'Australia and the South Pacific Nuclear Free Zone Treaty: A Reinterpretation', 567–83.

42 Matthew Jordan, 'Arms Control and Disarmament', in James Cotton and David Lee, eds, *Australia and the United Nations* (Canberra: Commonwealth of Australia, 2012), 282–3.

43 Lawrence Wittner, *Resisting the Bomb: A History of the World Nuclear Disarmament Movement, 1954–1970* (Stanford, CA: Stanford University Press, 1997), 206.

44 'Communique from Heads of Government Attending the South Pacific Forum Meeting, Cook Islands, 5–6 August 1985', *Backgrounder*, Annex, no. 491 (14 August 1985): A17–A18.

45 'Transcript of Prime Minister's Press Conference' (Rarotonga Forum Meeting, Rarotonga, 6 August 1985), 2.

46 Andrew Carr, *Winning the Peace: Australia's Campaign to Change the Asia Pacific* (Melbourne: Melbourne University Press, 2014), 158–9.

the outcome, the conclusion of the Rarotonga Treaty was seen by Australian officials as providing a critical boost to Australia's international profile as an activist middle power in the area of nuclear non-proliferation diplomacy. This point was acknowledged by the Department of Foreign Affairs in a cable to several Australian overseas posts on the eve of Australia's signature of the Treaty:

> [South Pacific] Forum governments appear to be unaware of the high level of international interest and, in most quarters, support which signature of the treaty [will] generate. There has, after all, been a rather severe and prolonged drought in the arms control field. The local peace movement in the South Pacific will make cynical, denigrating remarks but the international reaction can be expected to be overwhelmingly positive. The general international view will be that the South Pacific has made a contribution within its power to global stability and the prospects for disarmament, not that it ha[s] done something ineffectual and selfish [in] seeking to become a sanctuary.[47]

International reaction to the Rarotonga Treaty was generally favourable. This appears to have been due to two factors. The first, and probably the most significant, was the absence of any recent agreements in the field of nuclear arms control; SPNFZ was the first international nuclear arms control agreement struck since the initialling of SALT II by the Soviet Union and the United States in 1979. The second factor was that the Treaty of Rarotonga represented only the second regional nuclear free zone to be constituted in a populated area since the conclusion of the Treaty of Tlatelolco in 1967. Viewed from this perspective, the Rarotonga Treaty was seen as auspicious in promoting nuclear free zones as a means of bolstering the non-proliferation regime.[48]

Among the nuclear weapons states, reaction to the Rarotonga Treaty varied. While the Soviet Union's praise[49] followed China's expression of support for the Treaty earlier in 1985,[50] both Britain and France remained silent in the wake of the Rarotonga Forum meeting. Yet, the single most important reaction from the Hawke government's viewpoint was that of the United States. Although the public response of American officials was non-committal, privately the

47 Australian Department of Foreign Affairs, Outward Cablegram, 'SPNFZ: Handling at Rarotonga', O.CH302030 1848, 1 August 1985, NAA File: 277/17/1, Part: 16, 5.

48 On this point, see J. Goldblat and S. Lodgaard, 'The South Pacific Nuclear Free Zone: Variation on a Latin American Theme', *New Zealand International Review* 11, no. 3 (May/June 1986): 16–17.

49 J. Perrett, 'Nuclear Pact Wins Soviet Praise', *The Weekend Australian*, 28–29 September 1985.

50 *Arms Control Reporter* 4, no. 4 (April 1985): 456.B.13.

Reagan administration was uncomfortable with the outcome.[51] Despite the Hawke government's efforts to safeguard US nuclear transit rights during the negotiating process, and criticism that its SPNFZ policy was excessively geared towards American interests,[52] there is evidence that Washington remained critical of Australia's role in encouraging the SPNFZ process. During a visit to Washington in late August 1985, Minister for Defence Kim Beazley was informed that, notwithstanding the inclusion of permissive nuclear transit provisions in the Treaty, elements in the Departments of State and Defense believed that 'the establishment of the SPNFZ would make it progressively harder for US vessels to make port calls and transit through the region', and that 'the establishment of the zone was equivalent to giving a rod to the Russians to beat over the American back'.[53]

American endorsement of the Treaty would have served to strengthen the credibility of the government's active role in the SPNFZ process among conservative sections of the Australian electorate and lend support to the Labor government's repeated claim that the Rarotonga Treaty was compatible with Australia's alliance obligations. Indeed, US support for the Treaty appears to have been regarded by the Hawke government as a legitimate expectation in view of the Australian diplomatic capital expended in ensuring that the SPNFZ did not impinge upon American strategic operations in the South Pacific – in his Cabinet submission on the treaty, Foreign Minister Hayden even stated that it was 'designed to maintain the security advantages afforded to the South West Pacific through the ANZUS Treaty and the United States security presence in the region'.[54] Overall, the response from the United States did not amount to an endorsement of the Hawke government's approach – but the United States also did not make its reservations public. As in the case of the Joint Facilities, Australia had demonstrated through the SPNFZ that it would pursue and seek recognition of its own interests in regards to nuclear weapons and nuclear order. And yet, despite US reservations, that demonstration was also again carefully calibrated not to become a practical challenge to US nuclear policy and Australia's alliance with the United States.

---

51 For instance, see some of the non-attributable comments by administration officials in L. Sun, 'South Pacific Nations Declare Nuclear-Free Zone', *The Washington Post*, 8 August 1985.

52 The government's strongest Parliamentary critic along these lines was Independent Western Australian Senator Jo Vallentine who was elected in 1984 on the Nuclear Disarmament Party's ticket. For a sample of her views, see H. Conkey, 'N-Free Zone Plan "So Defective"', *The Canberra Times*, 29 July 1985.

53 Australian Department of Foreign Affairs, Inward Cablegram, 'SPNFZ', O.WH68884 1845, 30 August 1985, NAA File: 277/17/1, 3.

54 Cabinet Submission No. 2806 and Cabinet Decision No. 5512/ID and No. 5563, folio 9, NAA A14039, 2806.

Entrenching the 'Grand Bargain': Uranium Exports and ASTEC Review

One of the main nuclear policy challenges for the Hawke government was addressing the strong discontent within the Labor Party over Australia's uranium exports: The Labor platform of 1982 had stated that 'Labor is committed to reversing the direction of nuclear developments worldwide,' and that 'an Australian Labor government will adopt the process of phasing out our involvement in uranium mining, and certain conditions will be imposed upon the producers in accordance with this policy and we will utilise our right to apply trade embargos to ensure these conditions are met.'[55] Upon coming to office, Hawke announced a suspension of exports to France because of that country's nuclear weapons tests in the South Pacific.

On 31 October 1983, Cabinet then considered several issues in relation to Australia's uranium policy, all of which aimed at reducing the political saliency of the issue for the new government. On the one hand, it abolished the Uranium Advisory Council, which had provided a forum for community groups to comment on nuclear mining and uranium export.[56] On the other hand, in line with the party platform it refused further government facilitation of the still-ongoing UEGA study into the establishment of a uranium enrichment plant in Australia, which essentially stalled any further progress because no classified information could now be exchanged between the commercial partners.[57]

The most difficult issue was however the future of uranium mining and exports, and how to interpret the Labor Party's platform's commitment not to allow new uranium mines and to phase out Australia's involvement in the uranium industry. Cabinet considered options from the immediate cessation of all exports, to the development of the industry without restrictions. Political and economic considerations were in direct conflict, as the former option would reduce export revenue by $3 billion over 10 years, while the latter was estimated to increase revenue by $7 billion. In regards to normative concerns, the Minister for Resources and Energy, Peter Walsh, noted in his submission to Cabinet that there is 'concern that Australia's influence in these international efforts [safeguards] depends on it remaining a significant supplier or uranium'.[58] Cabinet in the end decided to allow development of the Roxby mine, and for all three existing uranium mines to continue operating and negotiate new contracts. However, approval for new exports was placed on halt until the Australian Science and Technology Council (ASTEC), an independent statutory body, concluded an inquiry into Australia's role in the nuclear fuel cycle. All future uranium exports, including under existing contracts, would then be subject to the conditions the government would put in place after

55 See 'ALP 1982 Part Platform: Uranium', NAA A11116, CA176 Part 1.

56 Cabinet Decision No. 2355, 31 October 1983, NAA A13977, 244.

57 Ibid., 246.

58 Cabinet Submission No. 385, 31 October 1983, NAA A13977, 385.

considering the inquiry's report.[59] ASTEC was assigned the task of examining existing international safeguards arrangements, Australia's role in strengthening the non-proliferation regime as a nuclear fuel exporter and questions relating to nuclear waste disposal. Through the review, the Hawke government thus began to shift Labor's uranium export policy,[60] as the terms of reference reflected the basic principles of Fraser's 'grand bargain'.

The philosophy of the ASTEC Report embodied a strong commitment to Australia's continuing participation in the international nuclear fuel cycle through the export of uranium. In stark contrast to the equivocal tone of the Ranger Inquiry's recommendations, the Report's principal conclusion was that '[t]he strength and effectiveness of Australia's voice in bilateral and multilateral forums and our ability to influence international actions designed to prevent the spread of nuclear weapons are *dependent* on our involvement in the nuclear fuel cycle.'[61] As was the case with the Ranger Inquiry, central to the ASTEC Report's conclusions was an evaluation of the state of international safeguards arrangements. While stressing the role of diplomatic measures in containing the nuclear weapons capacity of 'threshold' states, including pressure to accede to the NPT, the Report argued that these measures needed to be augmented by the 'acceptance of all treaty parties not to supply nuclear items to any non-nuclear weapon states for peaceful purposes unless those states' entire nuclear industries [are] subject at all times to IAEA safeguards'.[62] To reduce discrimination against NPT parties in the international nuclear marketplace, the Report recommended that Australia should actively promote full-scope safeguards as the uniform requirement for exports by all supplier countries.[63] The Report proposed that 'Australia encourage [the] further development of international guidelines … for the supply of nuclear items with a view to ensuring that countries which are parties to the Non-Proliferation Treaty or a treaty of similar coverage are advantaged'.[64] In addition, it recommended that Australia stay engaged in international technical cooperation on safeguards and nuclear technology.

In his tabling statement in parliament, Prime Minister Hawke referred to the report's remarks that progress towards disarmament was necessary to maintain the viability of the non-proliferation regime in general, especially in light of the 1985 decision on the extension of the NPT. This allowed him to draw direct links to the Australian support for the SPNFZ – continuing

59 Cabinet Decision No. 2352, 31 October 1983, NAA A13977, 385.

60 Sheila Hayden, 'Hawke Sees Approval of Uranium Developments As Step Towards Modifying ALP U308 Policy', *Nuclear Fuel* 8, no. 24 (21 November 1983): 2–3.

61 A.O. Slatyer, *Australia's Role in the Nuclear Fuel Cycle: A Report to the Prime Minister by the Australian Science and Technology Council (ASTEC)* (Canberra: Australian Government Publishing Service, 1984), 19. Emphasis added.

62 Slatyer, *Australia's Role in the Nuclear Fuel Cycle*, 9.

63 Ibid.

64 Ibid., 11.

a basic argument about the link between arms control and non-proliferation already made by the Fraser government, but also leveraging the SPNZ initiative in the context of the domestic uranium export debate.[65] When it considered the detailed recommendations, government rejected calls to allow additional mining and fuel cycle involvement, but accepted the large majority, which related to proposals to work internationally to strengthen non-proliferation norms, to detailed improvements to Australia's bilateral safeguards agreements, especially in relation to trans-shipment of materials, and to continued involvement in nuclear research. Importantly, government accepted that it would:

> maintain and strengthen its credentials [for appointment to the IAEA Board of Governors] through supply of Australian uranium for peaceful non-explosive purposes, and by continuing research and development programs in appropriate fields of nuclear technology …[66]

The Fraser government's 'grand bargain' was thus confirmed as the bipartisan orthodoxy of Australia's nuclear policy.

What remained was the special case of France. In September 1984, Cabinet had confirmed suspension of exports of uranium to France, in accordance with the Labor Party platform. The formal decision to suspend exports remained linked to the question of the nuclear tests, however, rather than France's refusal to sign the NPT, or the absence of full-scope safeguards over the country's military nuclear complex.[67] In fact, the government accepted that the suspension might even harm its non-proliferation policy, as '[w]e could indirectly be seen as lending support to those countries which argue that assurance of supply (however achieved) is more important than non-proliferation. This in turn would have implications for the 1985 NPT Review conference process.'[68]

In 1986, the government then overturned the ban on largely economic grounds. As the Minister for Resources Gareth Evans explained in Parliament, '[t]he policy has had a non-existent political impact in France' but at the same time, '[t]he budgetary implications of the existing embargo policy were considerable.' Allowing exports would provide $66 million, and 'not many other avenues offered that saving in a way that was not immediately hurtful to sections of the Australian community'. Moreover, these funds would 'go straight into our current account and, as such, be an important contributor in alleviating the larger balance of payments problem we obviously have'. With promised

65 Rt Hon. Robert Hawke, 'Prime Minister: Tabling Statement on ASTEC Report on Australia's Role in the Nuclear Fuel Cycle', Department of Prime Minister and Cabinet, 31 May 1984, available at http://pmtranscripts.dpmc.gov.au/browse.php?did=6400 [accessed 20 July 2014].

66 Cabinet Submission No. 2770, 6 May 1985, NAA A14038, 2770.

67 Cabinet Submission No. 997, 3 September 1984, NAA A13977, 997.

68 Ibid.

revenue of $200 million per year by the early 1990s, he went on, '[i]n our present economic climate, it is a considerable luxury to forgo an economic advantage for the nation of that magnitude.'[69]

The Continuing Quest for Full-Scope Safeguards for Non-NPT States

The issue of uranium mining and exports underscored sharp internal divisions in the Labor Party and provoked criticism from anti-nuclear sections of the electorate. At the 1984 Labor Party National Conference, a clause was inserted in the 'Uranium' section of the party's platform urging the government to:

> pursu[e] agreement at the 1985 NPT Review Conference that no NPT member State shall ever supply nuclear materials, equipment and technology to any non-nuclear weapons state which ... does not accept full-scope safeguards.[70]

As one of a host of specific arms control measures endorsed in the 'Uranium' section of Labor's platform (including the objective of a CTB treaty and the creation of regional nuclear-free zones in the Asia-Pacific), the full-scope objective was part of the broader trade-off between collective acceptance (however finely balanced) by the Party of a policy of restricted uranium exports and the adoption of an active nuclear policy.

In line with the recommendations of the ASTEC Report and the Labor Party platform, the Hawke government pursued the objective of international agreement to full-scope safeguards as a pre-condition for nuclear exports at the 1985 and 1990 NPT Review Conferences. However, the government realized that the international context for this would be challenging, and Foreign Minister Hayden submitted to Cabinet that 'the main threat to the [1985] Conference comes from some radical G77 who believe that their threats to damage the NPT would place pressure on the superpowers with respect to disarmament.' In this context, and in as good an illustration as any of the practical logic of the 'grand bargain', he wrote that conclusion of the safeguards agreement with Egypt was 'crucial' to work with that country at the conference.[71]

Despite active lobbying by Australian representatives at the NPT Review Conference, however, there was little support from leading nuclear suppliers to incorporate a full-scope commitment in the final declaration. Opposition to the provision of full-scope safeguards expressed at the 1980 NPT Review Conference was repeated in 1985 by key nuclear suppliers, including Switzerland, West Germany and France, while Japan, the United States and the United

69 *C.P.D.* (Senate), 20 August 1986.

70 Australian Labor Party, *Australian Labor Party Platform, Constitution and Rules as Approved by the 36th National Conference* (Canberra: National Secretariat, 1984), 178.

71 Cabinet Submission No. 2771, 16 April 1985, NAA A14039, 2771.

Kingdom remained lukewarm about any commitment.[72] Facing certain defeat of their proposal, Australian and Canadian representatives diluted the wording in a compromise formula that merely required nuclear exporters to 'take effective steps to achieve' the application of full-scope safeguards in countries receiving their exports.[73]

Between 1985 and 1990, there was no evidence that the NPT Review Conference recommendation on full-scope safeguards had any effect on the behaviour of exporting states. With the Nuclear Suppliers Group in abeyance, and the decisions by France and the Soviet Union to start negotiations on the export of nuclear technology to non-NPT member states operating unsafeguarded nuclear facilities (India and Pakistan), the prospects for tightening international nuclear export controls looked bleak in the lead-up to the 1990 NPT Review Conference.[74] Hopes for inserting a commitment to full-scope safeguards in the 1990 conference declaration were subdued because of the risk that states unhappy with the lack of progress towards disarmament by the nuclear powers would refuse to support a final declaration. Australia's strategy in 1990 was to persuade those countries that were lukewarm about the precondition of full-scope safeguards in 1985 to support a proposal that would make it difficult for opponents to maintain their position. Working with a range of like-minded countries, Australia helped draft a proposal that required full-scope safeguards be an essential precondition for nuclear exports.[75] A number of European suppliers registered strong reservations about the proposal, but this opposition then dissipated following Britain's defection after strong pressure from Australia and a newly reunified Germany.[76]

The full-scope requirement was adopted for inclusion in the final document, but the 1990 NPT Conference failed to reach consensus on a formal declaration overall. This diminished the force of the agreement reached at the conference to full-scope safeguards as a universal precondition. The Hawke government

72 See Australian Department of Foreign Affairs, *Third Review Conference of the Parties to the Treaty on the Non-Proliferation of Nuclear Weapons, Geneva, 27 August to 21 September 1985: Report of the Australian Delegation* (Canberra: Australian Government Publishing Service, 1985), 22.

73 Australian Department of Foreign Affairs, *Third Review Conference*, 22.

74 Harald Müller, David Fischer and Wolfgang Kötter, eds, *Nuclear Non-Proliferation and Global Order* (Oxford: Oxford University Press, 1994), 70–71.

75 Australian Department of Foreign Affairs, *Fourth Review Conference of the Parties to the Treaty on the Non-Proliferation of Nuclear Weapons, Geneva, 20 August to 15 September 1990: Report of the Australian Delegation* (Canberra: Australian Government Publishing Service, 1990), 67–8.

76 David Fischer and Harald Müller, 'The Fourth Review Conference of the Non-Proliferation Treaty', in Stockholm International Peace Research Institute, *World Armaments and Disarmament: SIPRI Yearbook 1991* (Oxford: Oxford University Press, 1991), 579.

was quick to claim victory, nonetheless, with Foreign Minister Gareth Evans noting that it 'set a new and historic benchmark for responsible behaviour in international nuclear trade'.[77] However, it was not until revelations of the rapid inroads that Iraq had been able to make into developing a nuclear weapons program that the Nuclear Suppliers Group committed itself to full-scope safeguards as a pre-condition for nuclear exports. There can be little doubt that Australia's efforts – in partnership with like-minded states – in getting the full-scope issue on the agenda of the 1985 and 1990 NPT conferences was a factor in encouraging eventual universal acceptance.[78] In the end, however, it was the end of the Cold War, realization of proliferation in Iraq and the consequent change in the non-proliferation policies of the major European countries with nuclear industry that brought about this long-sought Australian policy objective.

Reforming the Domestic Nuclear Institutions

Compared to the international regime, the Hawke government had much greater scope to refashion the domestic basis of Australia's non-proliferation commitments, and to review, revise and strengthen relevant acts in Australian domestic law. Its efforts to do so were partly due to the domestic political pressure from the left side of the political spectrum. In October 1985, the Labor Caucus created a special committee including Senator Evans, then Resources Minister, as well as Foreign Minister Bill Hayden and soon-to-be Defence Minister Kim Beazley, to 'consider the matter of legislative and other responses to the Democrats' Anti-Nuclear Bills'.[79]

In this context, the government's preference was to enshrine non-proliferation obligations in legislation even if that was not technically required. In relation to the provisions of the SPNFZ, for example, the government chose to create a new *South Pacific Nuclear Free Zone Treaty Act* to incorporate its obligations in Australian domestic law. It did so despite the risk that, in the words of Foreign Minister Hayden's submission to Cabinet, some 'Congressional and other circles [in the United States] may perceive the legislation in simplistic terms as a spreading of New Zealand's views on security issues with little inclination to look at the details of the Bill'.[80] While this and the more substantial changes discussed below streamlined Australia's bureaucratic arrangements and legal

77 Australian Department of Foreign Affairs and Trade, 'Nuclear Non-proliferation: Fourth Review Conference', *Minister for Foreign Affairs and Trade News Release*, no. M158 (17 September 1990).

78 Australian Department of Foreign Affairs and Trade, 'Nuclear Suppliers Group to Strengthen Information Exchange, Conditions of Supply', *Peace and Disarmament News* (March 1992), 14.

79 Cabinet Decision No. 7675, 2 June 1986, NAA A14039, 3960.

80 Ibid.

system in relation to nuclear matters, they also reflected a deeper desire by the Hawke government for a significant break with Australia's previous approach to nuclear policy.

When the Hawke government came to office in 1983, the *Atomic Energy Act 1953* was still the main legal basis for nuclear policy in Australia. The Act established the AAEC, created government controls over uranium deposits 'for the purposes of the defence of the Commonwealth' and established significant penalties for the unauthorized release of classified information relating to nuclear matters.[81] It had been amended several times over the years, especially under the Fraser government, which had revised it to provide a uniform legal basis for the regulation of uranium mining (although it still had to concede to the states a right of veto over commercial mining operations).[82] In the context of the planned establishment of enrichment facilities in Australia, the Fraser government had commenced a review of the act that got underway in 1981, with a view to reduce the role of the AAEC and establish a framework for a private sector nuclear industry in Australia.[83]

To the Labor Party, the Act (and the AAEC) represented all those nuclear aspirations and policies it abhorred. The AAEC had been associated with strong advocacy of nuclear power for Australia, and its long-time leader, Sir Philip Baxter, had been one of the most prominent Australian bomb advocates during the 1960s. The Agency was therefore tarnished in the eyes of many in the Hawke government, and a number of senior ministers were keen to overhaul its role to reinforce a sharper distinction between what they saw as legitimate civil applications of nuclear technology and illegitimate dual-use application that had the potential to aid a weapons program.[84] Shortly after the Hawke government assumed office, it decided to terminate the AAEC's work on centrifuge research and development. The ongoing review of the act was now re-focused on reforming the nuclear research establishment and embedding non-proliferation safeguards in Australian domestic law.

The AAEC was disestablished and a new, Australian Nuclear Science and Technology Organization (ANSTO) created through its own, separate act. In the debate on the ANSTO Bill, Gareth Evans described the AAEC as representing:

> that kind of 1950s approach to the nuclear industry: Tremendous cold war orientation, great emphasis on security; a tremendous inward looking character

---

81 Commonwealth of Australia, *Atomic Energy Act 1953–1973*, available at http://www.comlaw.gov.au/Details/C2004C01926 [accessed 20 July 2014].

82 Cabinet Decision No. 8442 (Ad Hoc), 10 May 1979, NAA A12930, 155.

83 John Carlson, email to authors, 8 December 2014.

84 For discussion, see Anna Eugenia Binnie, *From Atomic Energy to Nuclear Science: A History of the Australian Atomic Energy Commission* (Doctor of Philosophy thesis, Macquarie University, 2003), chapter 8.

> about it; a tremendous sense of being engaged in work that was dangerous and secret, and all those things which bear very little resemblance to the reality that now exists.[85]

Nonetheless, the government was also keenly aware that nuclear technology research made valuable contributions to the non-proliferation regime. In his Cabinet submission on the ANSTO act, Minister Evans thus highlighted that the act:

> [would] confine the new ANSTO primarily to nuclear science research and development although there will still be a need for an ANSTO capability to provide expert technical advice to Government on a range of nuclear matters to assist in the development of Government policy, particularly on nuclear non-proliferation safeguards and export controls.[86]

While ANSTO was thus still to conduct nuclear research and provide advice to government, it did not any more have a mandate to promulgate commercial adoption of nuclear power, or a role in relation to defence purposes: the ANSTO Act specifically prohibited 'research or development into the design or production of nuclear weapons or other nuclear explosive devices'.[87] Defence purposes were now also explicitly ruled out in the revised Atomic Energy Act.[88] At the same time, its 'repressive security provisions including reversal of the onus of proof in certain circumstances', which 'have been of long standing concern to the Australian Labor Party', were also repealed.[89]

The second major legislative change related to the operation of the Australian Safeguards Office (ASO). After the conclusion of Australia's INFCIRC/153 agreement with the IAEA, the ASO had been created as part of the AAEC to implement the national system of accounting of nuclear materials. In 1978, the Fraser government had moved the office to the Department of National Development,[90] responding to perceptions of a conflict of interest as the AAEC increased its role in commercial uranium mining.[91] With the resumption of uranium exports, the ASO was then also administering Australia's bilateral safeguards agreements with countries importing Australian uranium. The application of safeguards in Australia itself however relied on Customs

85 *C.P.D.* (Senate), ANSTO Bill 1985, 5 December 1986, 3570.

86 Cabinet Decision No. 6151, 17 June 1985, NAA A14039, 2421.

87 Commonwealth of Australia, *Australian Nuclear Science and Technology Organisation Act 1987*, section 5(2).

88 Commonwealth of Australia, *Atomic Energy Act 1953–1973*.

89 Cabinet Decision No. 6151, 17 June 1985, NAA A14039, 2421.

90 Cabinet Decision No. 7119, 9 November 1978, NAA A13075, 7119.

91 Cabinet Decision No. 8442 (Ad Hoc), 10 May 1979, NAA A12930, 155.

regulations and voluntary cooperation of the holders of nuclear materials in Australia.

In 1985, the Hawke government decided to reform the legal basis of the administration of safeguards in Australia, as well as other controls on nuclear technology and materials that flowed from export control regimes and the (then signed but not yet ratified) Convention for the Physical Protection of Nuclear Material. Instead of using regulations under the Atomic Energy Act, it decided to establish a separate, dedicated Nuclear Non-Proliferation (Safeguards) Act that created a national system of permits for the possession and transport of nuclear materials, and the communication of relevant information (including in patents).[92] It also created associated offences and penalties, and provided a legal basis for the activity of IAEA inspectors in Australia. In addition, the Act established the Director of Safeguards as an independent statutory officer under the act,[93] who would head the ASO (which had by then become part of the Department of Resources and Energy). At the time, Gareth Evans referred to the new Act as a 'unique piece of legislation. I do not think there is any equivalent for it elsewhere in the world'.[94]

Neither the *SPNFZ Treaty Act* nor the *Nuclear Non-Proliferation (Safeguards) Act* was strictly necessary at the time to implement Australia's obligations. However, the *SPNFZ Treaty Act* established a precedent that the Coalition government would follow with the dedicated *Comprehensive Nuclear Test Ban Treaty Act 1998*, and the offences it created in domestic law. The *Nuclear Non-Proliferation (Safeguards) Act* provided a good basis for the implementation of international agreements and conventions on strengthening nuclear safety in following decades.[95] The permits system it established would also ease Australia's adoption and implementation of the IAEA's Additional Protocol, as one of the very first countries to do so, in the 1990s. Hence, the Hawke government's transformation of Australia's nuclear acts and institutions in the 1980s resulted in an arrangement for safeguards that would turn out to be international best practice, and have a lasting influence on Australia's approach to implementing non-proliferation commitments.[96]

---

92 Commonwealth of Australia, *Nuclear Non-Proliferation (Safeguards) (Consequential Amendments) Bill 1987: Explanatory Memorandum*, available at http://www.austlii.edu.au/au/legis/cth/bill_em/nnab1987614/memo_0.html [accessed 20 July 2014].

93 Cabinet Decision (Amended) No. 6152, 17 June 1985, NAA A14039, 2422.

94 *C.P.D.* (Senate), 'Nuclear Non-Proliferation (Safeguards) Bill 1986', 18 February 1987, 209.

95 U.S. Department of State, *Treaty on the Non-Proliferation of Nuclear Weapons: U.S. Delegation to the 2010 Nuclear Nonproliferation Treaty Review Conference*, available at http://www.state.gov/documents/organization/141503.pdf [accessed 10 August 2014].

96 For an overview, see Nuclear Energy Agency, *Nuclear Legislation in OECD and NEA Countries: Australia* (Paris: OECD, 2008).

Conclusion

As this chapter has shown, the Hawke government consolidated the grand bargain and effectively made it a bipartisan undertaking. Yet when Hawke became Prime Minister this was certainly not a foregone conclusion as the new government had to grapple with the internal factional divides over core nuclear policy issues, such as the extent of Australia's involvement in the nuclear fuel cycle and the extent of Australia's participation in the ANZUS alliance (including the Joint Facilities), that had bedevilled the Labor Party since the late 1970s. Ultimately the Hawke government built on the policy framework established by the previous Fraser government. The key to this was its judgment that the core bargain struck by Fraser – that Australia's capacity to positively influence the direction of international non-proliferation efforts hinged on Australia remaining a significant supplier of uranium and maintaining appropriate nuclear science capacities – would place it in the best possible position to reconcile the country's interests across the strategic, economic and normative domains of nuclear policy.

Central to this task was the position the Hawke government took with respect to the vexed issue (at least from Labor's perspective) of uranium mining and export. By 1984 the government had adopted a policy of restricted uranium mining (via its Three Mines policy) coupled with a commitment to actively engage in arms control and nuclear disarmament initiatives, and seek the adoption of full scope nuclear safeguards as a precondition for nuclear trade. While this addressed anti-nuclear tendencies within the party, it nonetheless created a dilemma for the government with respect to the strategic domain of nuclear policy. The commitment to diplomatically push for arms control and nuclear disarmament in particular conflicted with Hawke's agenda to establish the party's credibility vis-à-vis national strategic and defence policy and, in particular, to reassure Washington of its commitment to the ANZUS alliance in a global strategic environment characterized by heightened US-Soviet tensions and arms racing. This dilemma played out most fully during Australia's involvement in the ultimately successful negotiation of the SPNFZ in 1986 whereby the Hawke government pursued a course of action consistent with the Prime Minister's stated desire for an 'independent and self-respecting' foreign policy despite the reservations of the Reagan administration. But while there were occasional tensions over this as well as the MX tests and Joint Facilities, the Hawke government was also careful to seek outcomes on these issues that would neither threaten basic underpinnings of US nuclear posture, nor lead to a rift within the broader Western alliance.

Chapter 5
New Opportunities and New Dangers: The Keating Government (1991–96)

Paul Keating became Prime Minister in December 1991 after winning a challenge against Bob Hawke in the Labor parliamentary caucus. Gareth Evans continued to serve as the Foreign Minister until the end of the Keating government. Before 1988, Evans had had responsibility for nuclear safeguards as Minister for Resources, and this provided an important element of continuity on arms control and non-proliferation across the Hawke and Keating governments. Although he shared his predecessor Bill Hayden's interest in arms control, Evans was less focused on nuclear issues specifically and devoted more effort to seeing the Chemical Weapons Convention to its conclusion in 1989.

The end of the Cold War provided the Keating government with a different, and generally far more favourable, environment for arms control than had been the case under Hawke.[1] In July 1991, the United States and the Soviet Union signed the Strategic Arms Reduction Treaty (START), which entrenched significant reductions in Cold War nuclear arsenals.[2] A few days after Keating became Prime Minister, the Soviet Union itself dissolved. Rapid progress was made towards universality of the NPT, and more countries became members between 1990 and 1994 (including France, China and South Africa) than during the whole of the 1980s, and only nine countries remained outside the treaty by the time the Keating government left office. The vexed question of full-scope safeguards as a condition for nuclear exports to non-NPT member states rapidly lost salience during the Keating years, not least because the Nuclear Suppliers Group finally agreed to adopt this condition in 1992.

---

1 For this reason, and because of the political continuity between the two governments, this chapter will thus include discussion of Australian policy beginning in 1989, from the time the late Hawke government began to adjust to the end of the Cold War.

2 Under the first Strategic Arms Reduction Treaty (START) signed in June 1991, the Soviet Union and the United States agreed to limit their nuclear warhead numbers to a ceiling of 6,000 and place limits on their deployed nuclear delivery systems. In 1993, under the START II agreement, Russia and the United States agreed to reduce their warhead numbers to 3,000–3,500 by 2003.

Also in 1991 the first Gulf War demonstrated the danger of the further proliferation of weapons of mass destruction to regional countries. After the war, international inspectors would uncover evidence of a large clandestine nuclear program in Iraq in violation of its NPT and safeguards commitments.[3] The same year, the IAEA was tasked with verifying the initial declaration under the South African comprehensive safeguards agreement, which would be complicated by South Africa's revelation in 1993 that it had manufactured nuclear warheads during the 1980s. Moreover, in 1993 North Korea's initial safeguards declaration was found to be false, which triggered a major crisis on the Korean peninsula bringing both sides to the brink of war. The IAEA safeguards system thus entered a new era in the early 1990s, with an increasing focus on the need to verify not only the correctness, but also completeness of states' declarations.[4] Whereas international non-proliferation diplomacy during the Fraser and Hawke years had largely focused on the extension of the NPT regime, under Keating attention would shift increasingly to the effectiveness of the regime in dealing with nuclear 'cheats'.

Geostrategic Uncertainty and Extended Deterrence

Australia and New Zealand were the Western allies least affected by the end of the Cold War, as it had ceased to be a dominant factor in their immediate strategic neighbourhood after the late 1960s. Hence, there was a strong element of continuity in Australia's strategic policy between the second half of the 1980s and the first half of the 1990s, and between the Hawke and Keating governments. Nuclear (and conventional) force reductions of the United States and Soviet Union were welcomed as contributions to global security, and as a sign that a more ambitious arms control agenda might now be feasible. But as the central balance itself had already diminished as a direct concern for Australia during the 1980s, this did not directly translate into specific Australian security gains either.

As such, the end of the Cold War presented the Keating government with a paradox. On the one hand, the world and hence Australia was undoubtedly safer after the end of the superpower confrontation. On the other hand, Cold

3 On Iraq's nuclear deception, see David Kay, 'Denial and Deception Practices of WMD Proliferators: Iraq and Beyond', *Washington Quarterly* 18, no. 1 (Winter 1995): 85–105.

4 See Laura Rockwood, 'Safeguards and Nonproliferation: The First Half-Century from a Legal Perspective', *Journal of Nuclear Materials Management* 35, no. 4 (2007): 79–93; Jonathan B. Tucker, 'Monitoring and Verification in a Non-Cooperative Environment: Lessons from the U.N. Experience in Iraq', *The Nonproliferation Review* 3, no. 3 (Spring-Summer 1996): 1–14.

War confrontation had also cemented great power relationships in Asia that had been the source of great stability from the late 1970s, and which now began to rapidly loosen. A classified update to Australia's strategic policy, endorsed by the government in 1989, stated that: 'Change in the familiar global order will mean greater strategic uncertainty elsewhere, including in Australia's region. A generally safer world does not necessarily mean a more tranquil region.'[5] What the sources of instability would be, however, was also not yet clear. The 1989 assessment's section on 'regional change' focused on domestic instability in the region, but also stated that: 'The prospect that competition between India and Pakistan and China could lead to further nuclear proliferation in South Asia is a concern for all countries of the region, including Australia.'[6]

Also in 1989, Foreign Minister Gareth Evans' ministerial statement on *Australia's Regional Security* highlighted that in the future, a range of major powers would have greater influence, including the United States, the USSR, Japan, the European Community, India and – with most caveats – China.[7] For the Keating government, following the demise of the Soviet Union, uncertainties about the future relationships and importance of the great powers in Asia became a central consideration, not least because US policy itself was 'in transition', as the 1993 *Strategic Review* put it. It stated that 'the United States' overseas commitments – and its underlying national security strategy – are still evolving',[8] which hinted at the concerns held by Australia and other US allies in Asia at the time about the reluctance of the Bush and Clinton administrations to clearly define America's commitment to the East Asian region.[9] Around the same time, the strategic implications of China's economic rise – and political stability despite the 1989 student protests – came to be appreciated as a major factor in Australia.[10]

---

5 Australian Department of Defence, *Australia's Strategic Planning in the 1990s* (Canberra: Australian Government Publishing Service, 1992), para. 1.6. The 1992 document is an unclassified version of the document endorsed by government in 1989.

6 Australian Department of Defence, *Australia's Strategic Planning in the 1990s*, para. 1.8.

7 Gareth Evans, *Australia's Regional Security* (Canberra: Department of Foreign Affairs and Trade, 1989), 4–7.

8 Australian Department of Defence, *Strategic Review 1993* (Canberra: Australian Government Publishing Service, 1993), para. 1.16.

9 Roger Bell, 'Reassessed: Australia's Relationship with the United States', in James Cotton and John Ravenhill, eds, *Seeking Asia Engagement: Australia in World Affairs, 1991–95* (Melbourne: Oxford University Press, 1997), 221. More specifically of concern to Australia had been comments in 1989 by US Vice President Dan Quayle that Australia could not automatically expect US assistance in a conflict with Indonesia. Bruce Jones, 'ANZUS: Is it Worth the Paper It's Written On?', *The Sydney Morning Herald*, 30 April 1989.

10 Bell, 'Reassessed: Australia's Relationship with the United States', 184–5.

In its 1994 Defence White Paper, the Keating government confirmed the basic strategic policy settings of the 1987 White Paper – but it also drew a direct link between Australia's security and developments in Asia, stating that:

> As long as the threat of global war remains low ... the likelihood of armed attack on Australia will depend on strategic developments in Asia and the Pacific, and particularly in Asia itself ... the relative peace in Asia may not last. The pattern of stable strategic relationships which has underpinned Asia's security in recent years is changing.[11]

In this context, it highlighted that:

> Much will depend on the policies of the major Asian powers – Japan, China and India – and on their relationships with one another and with other countries in the region. Perceptions of China's continuing economic growth and increasing political and military influence will be important, as will Japan's response to such factors as diminished US concern about Russia, China's approach to the region and the possible reunification of Korea.[12]

A key role in this regard would be played by the United States, whose policies and commitments not only helped to manage regional crises, but also influenced the policies of regional powers in general, and of Japan in particular:

> Whether more of Japan's strategic potential translates into military power, or its defence posture changes in other ways, will depend on how it perceives the capabilities and dispositions of its neighbours, and especially on Japan's confidence in its security alliance with the United States. That in turn is likely to depend on developments in the Korean peninsula.[13]

In 1995, the *United States Security Strategy for the East Asia–Pacific Region* (or 'Nye Report') committed to halting reductions in US forward-based forces at 100,000 personnel, thus signalling an enduring US commitment to security in Asia.[14] This was confirmed in a dramatic fashion in March 1996, when the Clinton administration dispatched two carrier battle groups to the vicinity of Taiwan to help counter Chinese threats against the island in the lead-up to Taiwan's presidential election.[15] Increasing economic integration had not prevented the

11 Australian Department of Defence, *Defending Australia: Defence White Paper 1994* (Canberra: Australian Government Publishing Service, 1994), 7.

12 Ibid., 8.

13 Ibid., 9–10.

14 Office of International Security Affairs, *United States Security Strategy for the East Asia–Pacific Region* (Washington, DC: Department of Defense, 1995).

15 R.S. Ross, 'The 1995–96 Taiwan Strait Confrontation: Coercion, Credibility, and the Use of Force', *International Security* 25, no. 2 (2000): 87–123.

crisis, and to the surprise of many, including in the United States and China, the two nuclear-armed countries had demonstrated they were ultimately willing to go to war with each other over the geopolitical shape of the new Asia.[16]

Hence, even though the 'central nuclear balance' had disappeared with the Soviet Union, the importance of US nuclear policies had not. Although they were not directed at a specific adversary in the way they were during the Cold War, US deterrence guarantees in Asia arguably increased in relevance to Australia during the early 1990s, because they were central to the symbolism and substance of US alliance commitments in the region, and hence to the strategic orientation of US allies. How Japan, South Korea or Taiwan would react to actual or proposed changes in US nuclear capabilities and policies would from now on be an essential consideration as Australia evaluated the role of nuclear weapons in international affairs.

In this context, it was only under the Keating government that Australia also began to pay public attention to the nuclear 'umbrella' that the ANZUS alliance provided to Australia itself. Anxiety about the vulnerability of the 'Joint Facilities' to Soviet nuclear attack during the Cold War in practice had been concerns about a Soviet attack on the United States that happened to take place on Australian soil, and therefore strictly speaking not an extended deterrence issue in which Australia would rely on the United States to come to its defence. With the end of the Cold War, however, Australian government statements began to reflect increasing appreciation of the US nuclear security guarantee implicit in the ANZUS alliance, beginning with the 1993 *Strategic Review* that stated, in a passage discussing the proliferation of weapons of mass destruction, that 'Australia continues to depend on the United States for security against any future nuclear threat.'[17] The 1994 White Paper for the first time referred to 'extended deterrence' as the strategy through which this dependence manifested itself. It stated that, on the one hand, it was 'hard to envisage the circumstances in which Australia could be threatened by nuclear weapons', but on the other hand, Australia 'will continue to rely on the extended deterrence of the US nuclear capability to deter any nuclear threat or attack'.[18] Hence, Australia would 'support the maintenance by the United States of a nuclear capability adequate to ensure that it can deter nuclear threats against allies like Australia'.[19]

Washington has endorsed the content and direction of this and subsequent Australian White Papers, and Australian officials have noted publicly and

---

16 J. Garver, *Face-Off: China, the United States, and Taiwan's Democratization* (Seattle: University of Washington Press, 1997), 3–12.

17 Australian Department of Defence, *Strategic Review 1993* (Canberra: Australian Government Publishing Service, 1993), 7.

18 Australian Department of Defence, *Defending Australia: Defence White Paper 1994*, 96.

19 Ibid.

privately that draft passages regarding US extended deterrence had been cleared with US officials before use in Australian strategic guidance documents. Former Deputy Secretary of the Department of Defence, Hugh White, stated that the initial insertion in the 1993 *Strategic Review* of a reference to the US nuclear umbrella (which was subsequently incorporated in the 1994 White Paper) was discussed informally with mid-level Pentagon officials who indicated 'they would get back to us if they had a problem'.[20]

Unlike the US alliances with Japan and South Korea, however, in relation to which the Clinton, Bush and Obama administrations have all publicly referred to extended nuclear deterrence commitments, AUSMIN communiqués or other joint declarations with Australia have not explicitly confirmed US nuclear guarantees to Australia. Australia became more similar to other US allies in publicly acknowledging the 'nuclear umbrella' through the comments in the 1994 White Paper (and subsequent documents under later governments). Yet, the 1994 White Paper's passages continued to reflect the more limited understanding of the role of nuclear weapons – i.e. to counter only other nuclear threats and not conventional ones – that had come to characterize Australian thinking on nuclear weapons during the Hawke years. In this, the Australian statements differed substantially from US policy and, for example, the two NATO *Strategic Concepts* of 1991 and 1999, which both contained the following paragraph (with only minor differences):

> To protect peace and to prevent war or any kind of coercion, the Alliance will maintain for the foreseeable future an appropriate mix of nuclear and conventional forces based in Europe and kept up to date where necessary … Taking into account the diversity of risks with which the Alliance could be faced, it must maintain the forces necessary to [ensure credible deterrence and to (1999)] provide a wide range of conventional response options. But the Alliance's conventional forces alone cannot ensure [the prevention of war (1991)/credible deterrence (1999)]. Nuclear weapons make a unique contribution in rendering the risks of … aggression … incalculable and unacceptable. Thus, they remain essential to preserve peace.[21]

20 Interview with Hugh White, Canberra, 23 May 2012 (quoted in Andrew O'Neil, *Asia, the US and Extended Nuclear Deterrence: Atomic Umbrellas in the Twenty-First Century* [London and New York: Routledge, 2013], 113). Since 1994, the process of consultation between Canberra and Washington on Australian statements regarding extended nuclear deterrence has reportedly become more formal and structured. Lyon, 'Australia: Back to the Future?', 436–7.

21 North Atlantic Council, *The Alliance's New Strategic Concept* (Brussels: NATO, approved 8 November 1991), para. 38, available at http://www.nato.int/cps/en/natolive/official_texts_27433.htm [accessed 8 August 2013]; North Atlantic Council, *The Alliance's Strategic Concept* (Brussels: NATO, approved 24 April 1999).

Resistance to calls for a 'sole purpose' doctrine continues to be an enduring feature of the nuclear policy of the United States and many of its allies. Hence, it is perhaps less surprising than it may otherwise appear that Australian statements on US extended deterrence have never received public endorsement from the United States.[22] In regards to the 'sole purpose' doctrine implied in the 1994 White Paper, Australian and US policy differed substantially – a difference that would have come to light had extended deterrence commitments been incorporated into a joint US-Australian document.

Southeast Asia, Counterproliferation and Missile Defence

One element of continuity in Australian nuclear policy under the Keating government was that Australia remained particularly focused on preventing the spread of nuclear weapons into its immediate neighbourhood. The 1989 *Australian Strategic Policy* document had stated that: 'Although the lead-times for the development of weapons of mass destruction by regional countries are generally long, we need to continue our efforts, especially in the diplomatic field, to prevent their proliferation.'[23] Foreign Minister Gareth Evans' 1989 statement on *Australia's Regional Security* highlighted that Australia's ability to influence international developments was conditioned by geographic distance;[24] and that

> the introduction of various 'exotic' weapons capabilities such as chemical weapons or nuclear weapons, particularly if accompanied by the acquisition of ballistic missiles, would have a dramatic effect on our security environment. Neither is impossible.[25]

Therefore, it went on to state:

> Keeping these new categories of weapons out of the region will have to be a high priority, requiring political and diplomatic intervention at appropriate times. It will also need to be a continuing focus of our arms control efforts. Our commitment to the prevention of the proliferation of nuclear and chemical

22 Which also makes the doubt expressed by, for example, Richard Tanter, over whether the nuclear umbrella applies to Australia at all, somewhat less pertinent. Richard Tanter, 'Just in Case: Extended Nuclear Deterrence in the Defence of Australia', *Pacific Focus* 26, no. 1 (2011): 113–36.

23 Australian Department of Defence, *Australia's Strategic Planning in the 1990s* (Canberra: Australian Government Publishing Service, 1992), para. 3.32.

24 Evans, *Australia's Regional Security*, paras 45–6.

25 Ibid., para. 66.

> weapons reflects not only a concern about developments around the globe, but more particularly an awareness of our regional security needs.[26]

In 1993, the Keating government's *Strategic Review* highlighted increased proliferation concerns in the post-Cold War era, and observed that '[n]uclear, chemical and biological weapons have not yet penetrated into South-East Asia, and one of Australia's prime strategic interests is to prevent any such penetration.'[27] This point was reinforced in a dedicated section on 'preventing proliferation', which explicitly linked Australia's regional security to global non-proliferation regimes, stating that '[g]lobal verification systems are the principal source of reassurance regarding proliferation in our nearer region,' and that 'we need to treat South-East Asian countries as natural dialogue partners on arms control initiatives.'[28]

However, this focus on preventing proliferation in Australia's immediate region introduced an additional element of difference between US and Australian policy in regards to the spread of nuclear weapons, and weapons of mass destruction in general. In the Fraser and Hawke years, Washington and Canberra both looked to the NPT, IAEA safeguards and export controls as the main tools to manage proliferation. Differences arose largely within the NPT framework, regarding the emphasis that should be placed on nuclear weapons states' Article VI commitments to work towards disarmament (even if only in symbolic terms in the context of the CTB negotiations). In the post-Cold War context, however, US policy began to complement this by an increasing emphasis on consequence management under the Clinton administration's 'Counterproliferation' initiative launched in December 1993.

The United States Department of Defense defined counterproliferation as 'the full range of military preparations and activities to reduce, and protect against' the threat of weapons of mass destruction and their means of delivery.[29] Initially this seemed to include technical efforts to enhance the protection of US forces, to enhance deterrence and to roll back proliferation where it occurred. But the initiative was far from uncontroversial even in the post-Gulf War context. The US military resented new expenditure in a fiscally challenging environment, while the arms control community criticized counterproliferation for reducing the role of diplomacy and the implicit inclusion of the pre-emptive

26 Ibid., para. 67.

27 Australian Department of Defence, *Strategic Review 1993*, para. 1.15. The same point was also reaffirmed in paragraph 2.15 of the same document.

28 Ibid., paras 2.16 and 2.21.

29 United States Counterproliferation Program Review Committee, *Report on Activities and Programs for Countering Proliferation and NBC Terrorism, Executive Summary* (Washington, DC: U.S. Government Printing Office, October 2001), 4, available at http://www.nti.org/media/pdfs/40_5.pdf?=1318280983 [accessed 20 July 2014].

use of military force.[30] In this context, the Clinton administration's policies on counterproliferation and non-proliferation became increasingly blurred: the 1994 Department of Defense (DoD) Report to Congress, for example, stated that '[a]ll of DoD's activities in the proliferation field – prevention and the new focus on protection – combine to form the set of activities called counterproliferation.'[31] Nonetheless, the term 'counterproliferation' continued to be used by the US Air Force to refer to new counterforce capabilities against WMD targets and, in general, came to be understood to refer to the development of military capabilities for deterring and, if necessary, defeating a WMD-armed regional adversary by the United States.[32]

The counterproliferation initiative was a reaction to the real threat to US forces, in the Middle East and in Korea in particular, from existing weapons of mass destruction in the hands of countries hostile to the US presence in those regions. Dealing with such 'rogue state' capabilities was the main non-proliferation concern for the United States in the post-Cold War context, and the global non-proliferation regimes (not only in relation to nuclear, but also chemical and biological weapons) provided a useful political context to manage this concern. For Australia, however, the situation was the reverse. Concerned with keeping weapons of mass destruction out of Southeast Asia, the integrity of global regimes that maintained this status was its highest priority, and the management of 'rogue states' a problem subservient to that objective. Hence, David Cooper finds in a systematic comparison of US and Australian policies of the 1990s, that '[i]n virtually every case where tradeoffs have existed between norm-building and capability-denial and/or consequence-management, Australia has opted to support the former.'[33]

This was particularly true for the politically charged counterproliferation initiative, as overt US preparations for defeating regional powers were hardly conducive to international negotiations with the NAM in the lead-up to the 1995 NPT Review and Extension Conference, which was charged with determining whether the Treaty would be extended indefinitely or for subsequent rolling periods. Hence, when US officials visited Canberra to discuss cooperation on counterproliferation in 1994, 'senior Australian Defence officials express[ed] polite lack of interest while DFAT officials voiced open hostility,'[34] and neither

30 Thomas G. Mahnken, 'A Critical Appraisal of the Defense Counterproliferation Initiative', *National Security Studies Quarterly* (Summer 1999), 91–3.

31 United States Secretary of Defense, *Annual Report to the President and the Congress* (Washington, DC: U.S. Government Printing Office, 1994), 36.

32 Harald Müller and Mitchell Reiss, 'Counter-Proliferation: Putting New Wine in Old Bottles', *The Washington Quarterly* 18, no. 2 (Spring 1995): 143–5.

33 David A. Cooper, *Competing Western Strategies Against the Proliferation of Weapons of Mass Destruction* (Westport, CT: Praeger, 2002), 186.

34 Ibid., 142.

the 1994 White Paper nor the Howard government's 1997 *Australian Strategic Policy* document would include any reference to the concept or substance of counterproliferation. However, where additional measures against horizontal proliferation had less detrimental impact on the global regimes, Australia did support its major ally. Hence, the Keating government followed the examples of the United States and Germany, and introduced 'catch-all' clauses to restrictions on the export of WMD-relevant dual-use goods and services in the *Weapons of Mass Destruction (Prevention of Proliferation) Act* of 1994.[35]

A partial exception to Australia's aversion of engaging in counterproliferation research was in the area of missile defence, despite the serious reservations over the Reagan SDI program that the Hawke government had shared with many other US allies. The 1994 White Paper's paragraph on the Joint Facilities acknowledged that their role was shifting to also include warning of regional missile attack, and noted that 'such cooperation will continue to be in Australia's interest'.[36] In 1995 Australia agreed on practical cooperation with the United States in relation to missile defence, focused on sensor data fusion. After an initial test in Woomera in October 1995, the new Howard government would then agree to the launch of four US test missiles off Northwest Australia, which simulated the radar cross section of short range ballistic missiles and were observed by the Australian Jindalee Over-the-Horizon-Radar (JORN).[37] US officials at the time maintained that JORN technology could, in the long term,

---

35 John Faulkner in his second reading speech on the bill in the Senate made explicit reference to the fact that Australia was following the US and German examples. See Editors, 'Use of Force and War: Weapons of Mass Destruction – Prevention of Proliferation – New Australian Legislation', *Australian Year Book on International Law* 26 (1995): 549.

36 Australian Department of Defence, *Defending Australia: Defence White Paper 1994* (Canberra: Australian Government Publishing Service, 1994), para. 9.14. Reflecting the difficulty of adjusting the public justification for the joint facilities that had evolved under the Hawke government to the post-Cold War period, the White Paper's comments were somewhat obtuse. The White Paper's comments on the Joint Facilities in this respect were rather imprecise. For example, it stated that 'Our participation in the Defence Support Program contributes to global security against surprise or accidental nuclear attack,' even though there was then not yet a missile defence system that could have acted on the warning provided by the system.

37 See the Statement of Brigadier General Emery, USAF, Deputy for Theatre and Air Missile Defense, Ballistic Missile Defense Organization, to the House National Security Committee, Subcommittees on Research & Development and procurement (Washington, DC, 19 March 1997), available at http://www.mda.mil/global/documents/pdf/ps_emery.pdf [accessed 20 July 2014]; 'DUNDEE', Global Security.org Website, http://www.globalsecurity.org/space/systems/dundee.htm (last modified 6 August 2003); Missile Defense Agency, Project DUNDEE Fact Sheet, August 1997, http://www.acq.osd.mil/bmdo/bmdolink/html/dndefs.html (last modified 6 October 2003).

become a component of the American missile defence system.[38] However, both the Keating and Howard governments justified their involvement on missile defence research not because Australia saw a particular need for missile defence as such, but because Australia should be seen to be a cooperative ally, and the research also had other applications.[39]

IAEA Safeguards Reforms and the Additional Protocol

The discovery of the Iraqi clandestine nuclear program after the 1991 Gulf War provided a major impetus to international efforts to reform and strengthen the safeguards regime. In 1993, the IAEA's Standing Advisory Group on Safeguards Implementation (SAGSI) submitted a set of recommendations on safeguards reform. The resulting 'Program 93+2', timed to develop new arrangements for the 1995 NPT conference, included both measures that were within the existing legal powers of the IAEA inspectors under the INFCIRC/153 agreements ('Part 1'), and measures that would require additional legal arrangements ('Part 2').[40] The overall effect of the changes was a fundamental reorientation in focus, with less concern about the large nuclear infrastructure in Japan or Germany, especially their large plutonium reprocessing plants, and an increased focus on verifying the completeness of states' declarations to prevent proliferation through covert, dedicated facilities.[41] As part of this change, increased efficiency in the safeguards system was a crucial aspect, especially through remote monitoring and a shift to random inspections, as the IAEA inspection budget was largely static while the number of nuclear installations continued to increase.[42] In the end, the 'Part 2' measures of the 93+2 programme led to the development of the Additional Protocol, which gave the IAEA additional inspection powers, access to additional parts of the fuel cycle and early design information on nuclear facilities. In more recent years, the IAEA has developed 'integrated safeguards' in states that adopted both the traditional INFCIRC/153 and the Additional Protocol.[43]

The effect of these changes was that the legal basis, practical implementation and technology of safeguards returned to the centre-stage

38 Gregor Ferguson, 'Australia, U.S. Discuss SBIRS Collaboration Effort', *Defense News* 14, no. 42 (25 October 1999): 10.

39 Cooper, *Competing Western Strategies*, 185.

40 Dirk Schriefer, 'New Safeguards Measures: Initial Implementation and Experience', *IAEA Bulletin* 4 (1996): 7–10.

41 David A. Fischer, 'Viewpoint: New Directions and Tools for Strengthening IAEA Safeguards', *The Nonproliferation Review* (Winter 1996): 69–76.

42 Bruno Pellaud, 'Safeguards: The Evolving Picture', *IAEA Bulletin* 38, no. 4 (1996): 2–6.

43 Richard Hooper, 'The Changing Nature of Safeguards', *IAEA Bulletin* (June 2003): 7–11.

of non-proliferation policy. In the 1970s, Australia had participated in the development of the original safeguards approach, including through the 'Hexapartite project' on the inspection of centrifuge plants. Individual Australian experts from the ASO had been appointed as members of SAGSI for most of the time since the mid-1970s.[44] In 1977 the Fraser government had sought Australian representation on SAGSI,[45] as part of its broader efforts to help strengthen the IAEA safeguards system.[46] In 1980, it had funded a three-year member state support program to help develop safeguards technologies and approaches,[47] which the Hawke government had decided to continue for another three years in 1983.[48]

After 1986, however, dedicated funding had ended. In 1989–90, the ASO then recommenced a formal Australian Safeguards Assistance Program from its own resources, and a five-year program was funded by government from 1991.[49] In addition to providing analytical support on the development of new safeguards approaches, Australia agreed to work with the IAEA to develop and trial new safeguards measures in Australia from 1992.[50] Australia allowed additional access for IAEA inspectors, which led to a better understanding of the legal and practical implications of such inspections.[51] Remote monitoring equipment was installed at ANSTO's dry fuel storage facility at Lucas Heights as part of a two-year trial.[52] Australia allowed the IAEA limited access to stored components of the centrifuge program that had ended in 1983, and access to the buildings where the centrifuges once stood to test new environmental

44 Australian Safeguards and Non-Proliferation Office, *Annual Report 2011–2012* (Canberra: Australian Government Publishing Service, 2012), 71–2. For background on SAGSI, see John Carlson, *SAGSI: Its Role and Contribution to Safeguards Development* (Canberra: ASNO, 2007), available at http://www.dfat.gov.au/asno/publications/SAGSI_role_contribution_safeguards_dev.pdf [accessed 20 July 2014].

45 Cabinet Decision No. 3913, 21 September 1977, NAA A12909, 1666.

46 Cabinet Decision No. 2926, 16 May 1977, NAA A12909, 1232.

47 Cabinet Decision No. 11883 (Ad Hoc), 17 June 1980, NAA A12909, 4101.

48 Cabinet Decision No. 1544, 23 July 1983, NAA A13977, 143.

49 Australian Safeguards Office, *Annual Report 1993–1994* (Canberra: Australian Government Publishing Service, 1994), 29.

50 Australian Safeguards Office, *Annual Report of the Director of Safeguards 1994–1995* (Canberra: Australian Government Publishing Service, 1995), 10.

51 See J. Carlson, J. Bardsley and J. Hill, 'Trials in Australia related to the assurance of the absence of significant undeclared nuclear activities (IAEA-SM-333/132P)', in *International Nuclear Safeguards 1994: Vision for the Future* (Vienna: IAEA, 1994), 290–93.

52 Australian Nuclear Science and Technology Organisation, *Annual Report 1993–1994* (Lucas Heights: ANSTO), 61; D.L. Sorokowsi, K. Veevers and C.S. Johnson, 'Co-operative Activities of the Australian Safeguards Office and Sandia National Laboratories in Advanced Containment and Surveillance Technologies (IAEA-SM-333/107)', in *International Nuclear Safeguards 1994: Vision for the Future* (Vienna: IAEA, 1994), 631–40.

sampling techniques. Despite repeated renovations over 10 years, the IAEA was able to determine the former location of the centrifuges within the building, as well as the average and the maximum level of enrichment achieved.[53] In addition, Australia agreed to open the former nuclear weapons test area at Maralinga to the IAEA, should it ever seek to verify technologies to detect the remains of nuclear explosions.

The renewed focus on safeguards technology and implementation in the 1990s reinforced the importance of scientific expertise for nuclear policy and non-proliferation diplomacy. The ASO was therefore moved from Sydney to Canberra in 1990, so that it was co-located with the Federal government and could increase its role in the development of government policy.[54] As its involvement in IAEA safeguards development increased, the ASO was moved in 1994 from the Resources portfolio to the responsibility of the Minister for Foreign Affairs and Trade. At the same time, the Director of Safeguards assumed additional responsibilities for other arms control verification, and the office was therefore reconstituted as the Australian Safeguards and Non-Proliferation Office (ASNO) when it moved into the new portfolio.[55]

Since the early 1990s, the development of safeguards has remained an important element in the evolution of the non-proliferation system, especially in relation to the development by the IAEA of the 'state-level approach' to safeguards implementation in the late 2000s. ASNO and ANSTO expertise has allowed Australia to play a meaningful role in the technical aspect of the NPT regime, which remains an important aspect of Australia's non-proliferation activities. In 2002, for example, ANSTO's Accelerator Mass Spectrometry facility was the first such facility to be accredited by the IAEA for the analysis of environmental samples taken by IAEA inspectors,[56] and in 2012 the University of Western Australia also joined the IAEA Network of Analytical Laboratories. But even in relation to Australia's own policy initiatives, technical expertise provided an important means of support. The Canberra Commission's recommendation to tailor safeguards to the isotopic composition

---

53 Jodie Evans and Russell Leslie, 'Australia's gas centrifuge program and safeguards related research', ASNO (presentation, IAEA Technical Working Group on Enrichment, Vienna, 18–22 May 2005), cited in M.D. Laughter, *Profile of World Uranium Enrichment Programs, 2009*, ORNL/TM-2009/110 (Oak Ridge, TN: Oak Ridge National Laboratory, 2009).

54 Australian Safeguards and Non-Proliferation Office, *Australian Safeguards and Non-Proliferation Office Annual Report: 1998–1999* (Canberra: Australian Government Publishing Service, 1999), available at http://www.dfat.gov.au/asno/annual_report_9899/25_years.html [accessed 20 July 2014].

55 Ibid.

56 The Hon. Alexander Downer, MP, and the Hon. Peter McGuaran, MP, 'Boost for Australia's Nuclear Watchdog Role', joint media release, available at http://www.foreignminister.gov.au/releases/2002/fa065j_02.html [accessed 20 July 2014].

of plutonium stocks, for example, was recognized as an important contribution to the safeguards debate,[57] and ASNO was subsequently able to take up this issue as part of its technical safeguards support program activities.[58]

Global Nuclear Diplomacy and the Continuing Quest for a CTBT

During its more than four years in office, the Keating government devoted significant energy to maintaining Australia's engagement in international nuclear diplomacy. As had its predecessors, the Keating government sought to reinforce Australia's status as an activist middle power in international relations and limit the spread of nuclear capabilities globally, and in Asia particularly. This was reflected in the government's advocacy of the indefinite extension of the NPT in 1995. But it was also strongly evident in the detailed work it undertook on narrowing the scope of negotiations towards a CTBT, in its involvement in the debate over whether nuclear weapons should be illegal, and in the sponsorship of the Canberra Commission on the Elimination of Nuclear Weapons.

In each of these areas, the Keating government drew on a well-established set of policies laid down by earlier governments. This was particularly evident in relation to the CTBT. Achieving a comprehensive ban on nuclear testing represented the major preoccupation in Australian nuclear diplomacy until the conclusion of the CTBT in 1996. The Fraser and Hawke governments had argued consistently that a CTBT would present a critical long-term barrier for those states seeking to acquire nuclear weapons and serve to constrain nuclear force modernization programs among the nuclear powers. Australian governments also believed that a complete ban on nuclear testing would reinforce the norm of non-proliferation in international relations.

For the Keating government, its support for a CTBT was shaped by three factors. The first was the imperative of concluding a treaty before the commencement of the 1995 NPT Review and Extension Conference. Long identified as the single most urgent multilateral arms control measure, a CTBT was generally regarded as crucial in demonstrating the commitment of the nuclear powers to implementing Article 6 of the NPT. By assuaging the legitimate concerns of non-nuclear states that were critical concerning the slow progress in multilateral nuclear arms control, it was surmised that a CTBT treaty would clear an important obstacle for indefinite extension of the NPT in 1995.

The second factor was the potential role a CTBT could play in bringing the three smaller nuclear powers – Britain, China and France – into a broader

---

57 Pellaud, 'Safeguards: The Evolving Picture'.

58 Australian Safeguards and Non-Proliferation Office, *Australian Safeguards and Non-Proliferation Office Annual Report: 1998–1999* (Canberra: Australian Government Publishing Service, 1999), available at http://www.dfat.gov.au/asno/annual_report_9899/25_years.html [accessed 20 July 2014].

process of multilateral nuclear arms reductions after the Cold War. In contrast to the primary focus of previous governments on US-Soviet nuclear restraint, the Keating government linked a CTBT with promoting nuclear restraint on the part of all five nuclear weapons states. As Foreign Minister Gareth Evans observed in 1992:

> In the case of the UK, France and China, I think this is an appropriate occasion for the Australian Government to call on them … to follow the example of Russia and the United States and to start the process of implementing cuts to their nuclear arsenals. We do not accept that the smaller size of their arsenals in any way frees them from their continuing obligation to negotiate measures of nuclear disarmament.[59]

The third factor shaping the Keating government's support for a CTBT related to what Trevor Findlay has referred to as 'sunk costs'.[60] Since the early 1970s, many countries had invested significant political capital and diplomatic energy in efforts to attain a comprehensive ban on nuclear testing. As traditionally one of the most vigorous proponents of a CTBT, and having carried out technical work in the critical area of verification, Australia had a major stake in advocating a CTBT after the Cold War.

At the time of Paul Keating's accession to the Prime Ministership, the international climate for concluding a CTBT looked promising. As part of its unilateral nuclear force reductions announced in response to an earlier initiative by the Bush administration, in October 1991 the Soviet Union declared a 12-month moratorium on nuclear testing that extended the *de facto* one-year moratorium in place since the previous year.[61] In April 1992, France announced that it too would suspend its testing program for one year and linked any future extension of this moratorium to the other four nuclear powers refraining from testing.[62] In October the same year President Bush placed an immediate nine-month moratorium on American nuclear testing and a ban on all underground tests after 1996 on the condition no other state tested subsequently.[63] By the end of 1992, Britain, France, Russia and the United States had simultaneously suspended their nuclear testing programs for the first time since 1958.

---

59 *C.P.D.* (Senate), 17 June 1992, 3794.

60 Trevor Findlay, 'A Comprehensive Nuclear Test Ban: Post-Cold War Prospects', *Peace Research Centre Working Paper*, No. 122 (Canberra: Australian National University, September 1992), 12.

61 R. Haupt, 'Soviets Match Arms Cuts', *The Age*, 7 October 1991.

62 P. Jacobsen and M. Binyon, 'France Calls off Nuclear Testing', *The Times*, 9 April 1992.

63 D. Lockwood, 'US Begins Testing Moratorium, Ban on US Tests Set for 1996', *Arms Control Today* 22, no. 8 (October 1992): 32, 40.

Despite this auspicious climate, negotiations in the Conference on Disarmament (CD) between 1992 and 1994 moved slowly. In particular, there was disagreement about the scope of any ban on testing under the CTBT, as well as verification and entry-into-force provisions. During negotiations in 1994, Britain and France insisted that a provision be included in a CTBT that permitted nuclear testing for the purposes of 'safety and reliability'.[64] In contrast, Germany and Sweden insisted that any preparations for nuclear testing be prohibited.[65] Australia's position on scope was more generic, and the draft CTBT text it introduced into the CD in early 1994 was deliberately designed to avoid discussion of outstanding issues with the assumption being that these would be resolved as agreement converged on a 'rolling draft text'.[66] Yet by the end of 1994, divisions still remained over the question of scope, with the nuclear powers still referring to the possibility of 'sub-kiloton' testing and larger scale testing being allowed in 'exceptional circumstances' under a CTBT.[67]

In the broader context of CTBT negotiations, the year 1995 was a watershed in achieving substantive progress towards a treaty. Ironically, France's announcement in June 1995 that it would resume its underground nuclear testing program had a beneficial effect in contributing to the resolution of the single most important element in CTBT negotiations: the treaty's scope. This followed China's decision to undertake nuclear tests in the days following the conclusion of the NPT Review and Extension Conference. Unlike China, however, France had committed itself to a moratorium in 1992. The resumption of French nuclear testing provoked outrage in the Australian electorate, part of which was directed at the Keating government over what many saw as its timid initial response.[68] The government's subsequent endeavour to make up ground resulted in a decision to explicitly pursue the principle of zero-yield in the area of CTBT scope – i.e. no tests under any circumstances no matter how small or controlled.[69]

Aside from the high profile resumption of French nuclear testing, what other factors influenced the Keating government's decision to tighten up Australia's CTBT policy in mid-1995? An important factor related to shifting

64 *Arms Control Reporter* 13, no. 4 (April 1994): 608.B.301–6.

65 D. Lockwood, 'CD Negotiators Offer Proposals for Comprehensive Test Ban Treaty', *Arms Control Today* 24, no. 3 (1994): 26.

66 Editor, 'Australia Puts Forward Draft Treaty Elements', *Disarmament: Newsletter of the UN Centre for Disarmament Affairs* 12, no. 2 (1994): 3.

67 Eric Arnett, 'The Comprehensive Nuclear Test Ban', in *SIPRI Yearbook 1995: Armaments, Disarmament and International Security* (Oxford: Oxford University Press, 1995), 711–14.

68 Karen Middleton, 'Test Anger Misjudged, Says Evans', *The Age*, 28 June 1995.

69 Helen Trinca, 'US and Britain Argue French Will Sign Ban Treaty: Envoy', *The Australian*, 21 July 1995.

international opinion on the CTBT issue. During much of 1994 and early 1995, most non-nuclear states participating in multilateral negotiations (including Australia) had been content for the nuclear powers to agree among themselves on a CTBT yield threshold if it meant hastening progress towards a draft treaty.[70] However, with the Chinese and French decisions to carry out nuclear testing, speculation over the future of the American testing program, and British and Russian ambivalence over committing to a zero-yield scope, there was a souring of the negotiating atmosphere within the CD following the 1995 NPT conference. Some began to express the view that the 'timeliness' of a CTBT treaty had passed.[71]

Having tabled a loosely defined draft article on scope that received broad support within the CD, the Keating government realized that Australia had a vested interest in seeking agreement on what specific testing activities should be prohibited. To that end, it was seen as necessary to foster consensus among the nuclear powers to ensure agreement. Yet, it was seen as equally important to mollify the concerns of certain non-nuclear states, especially those in the NAM, that any CTBT treaty not be excessively lenient in permitting non-explosive testing activities. Uniting both of these considerations was a wider imperative of not letting negotiations fall apart entirely. Sensing a window of vulnerability for CTBT negotiations triggered by the intransigence of the nuclear powers, and sensing that the latter were on the defensive in response to international criticism of their actions, the Keating government decided that a literal zero-yield interpretation of nuclear testing activities offered the best way forward for promoting the conclusion of a CTBT. Australia's formal proposal on zero-yield in the Conference on Disarmament in 1994 was pivotal in encouraging all states – most importantly the nuclear powers – that a definition of scope under the CTBT that outlawed any form of testing was the framework for final stage treaty negotiations in 1995–96. The conclusion of the treaty negotiations then fell into the early days of the Howard government, which finally brought about an outcome that Australian governments since Fraser had sought to achieve.

The ICJ Case and the Canberra Commission

In the last months of the Labor government's time in office, there was a notable change in the rhetoric about nuclear disarmament. In October 1995 the Keating government declared that no longer would Australia remain merely an interested bystander in international debates on this particular objective. Until its electoral defeat in March 1996, the Keating government argued that if the proliferation

70 Editor, 'CTB Negotiations: Geneva Update', *Nuclear Proliferation News*, no. 29, 11 July 1995, 3.

71 Rebecca Johnson, 'Test Ban Treaty: Rearranging Deck Chairs on the Titanic', *Bulletin of the Atomic Scientists* 51, no. 5 (September/October 1995): 12.

of nuclear weapons was to be prevented, it could only be achieved in the context of a renewed international commitment to nuclear disarmament. As Foreign Minister Evans observed in early 1996, 'the really big objective, and the one that's absolutely crucial if we are to win on non-proliferation, is the ultimate elimination of nuclear weapons from the face of the world.'[72]

The Keating government displayed particular concern over the growing sense of resentment felt by many non-nuclear states following the indefinite extension of the NPT in 1995. Although the Labor government had been one of the most active supporters of the NPT's indefinite extension, it was worried by the reluctance of the nuclear powers to make real progress on arms control. As Keating's response to the French nuclear testing announcement would reveal, the primary reason for the government's anxiety was a belief that this reluctance had the potential to undermine the normative force of nuclear non-proliferation in international politics.

This perception was central in driving Australia's unprecedented activity on the issue of nuclear disarmament during the last six months of the Keating government. The government took two prominent initiatives during this time: Australia's involvement in the case before the International Court of Justice (ICJ) regarding the legality of the use or threatened use of nuclear weapons, and the creation of the Canberra Commission on the Elimination of Nuclear Weapons. While both were frequently invoked by the Keating government as evidence of Labor's genuine commitment to nuclear disarmament, electoral calculations and concern over nuclear proliferation were also major factors shaping each initiative.

In May 1993, the World Health Organization (WHO) had adopted a draft resolution requesting the ICJ to provide an advisory opinion on whether the use of nuclear weapons violated international law.[73] At that time, Australia's delegate to the WHO had vigorously opposed the WHO draft resolution, asserting that 'the legality of nuclear weapons was a political issue which did not fall into the arena of world health and it was therefore inappropriate for the [WHO] to consider it'.[74] In opposing the WHO resolution, Australia

---

72 Gareth Evans, 'Address by the Hon. Gareth Evans QC, Minister for Foreign Affairs, to the Rotary International World Understanding Month Dinner, Noble Park' (15 February 1996), 14, available at http://www.gevans.org/speeches2.html [accessed 12 June 2014].

73 Specifically, the WHO Resolution requested the ICJ to give an advisory opinion on the following question: 'In view of the health and environmental effects, would the use of nuclear weapons by a State in war or other armed conflict be a breach of its obligations under international law including the WHO Constitution?' *International Court of Justice Yearbook: 1993–1994*, no. 48 (The Hague, 1994), 208.

74 Cited in Nicholas Rostow, 'The World Health Organization, the International Court of Justice, and Nuclear Weapons', *Yale Journal of International Law* 20, no. 1 (Winter 1995): 155–6.

argued that the UN was the only appropriate forum via which any request to the ICJ regarding nuclear weapons legality could be conveyed.[75] The Keating government's formal position on the WHO Resolution was spelt out in more detail in mid-1994. Noting that although Australia remained 'very sympathetic to the motives' underlying the WHO request, Foreign Minister Evans outlined what the government regarded as the likely hazards resulting from any ICJ advisory opinion,[76] notably that anything but an unequivocal verdict would be very damaging for the disarmament cause.[77] Australia deposited a written submission with the ICJ in June 1994 recommending that the Court use its procedural discretion *not* to give an advisory opinion on the WHO request. However, the submission reserved Australia's right to make 'a substantive submission at a later time on the question of legality of use, should the court decide to consider issues of substance'.[78]

Towards the end of 1993, NAM states agreed to table a draft resolution at the UN along the lines of the earlier WHO request. However, facing furious resistance from Britain, France and the United States, and confronting divisions within its own ranks, NAM countries subsequently withdrew the draft without putting it to a vote.[79] While the nuclear powers had remained openly suspicious of the WHO initiative, their reaction to the NAM's draft UN resolution, which broadened the WHO resolution to include the threat of nuclear weapons use, was openly hostile.[80]

In December 1994 the NAM successfully re-tabled its 1993 UN draft resolution and achieved a narrow vote in favour. UN General Assembly Resolution 49/75K 'urgently' requested the ICJ to provide an advisory opinion on whether 'the threat or use of nuclear weapons [is] in any circumstances permitted under international law'.[81] Despite its stated preference that any request to the ICJ be conveyed via the UNGA, Australia abstained from voting on Resolution 49/75K.[82] Responding to criticism of Australia's abstention,

---

75 K. Dewes and R. Green, 'The World Court Project: How a Citizen Network Can Influence the United Nations', *Pacifica Review* 7, no. 2 (October/November 1995): 27.

76 *C.P.D.* (Senate), 2 June 1994, 1207–8. Emphasis added.

77 Gareth Evans, email to Stephan Frühling, 16 February 2015.

78 K. MacArthur, 'Australian Submission to the International Court of Justice on the Legality of the Use of Nuclear Weapons in Armed Conflict', *Peace and Disarmament News* (August 1994), 9.

79 M. Schapiro, 'Mutiny on the Nuclear Bounty', *The Nation*, 27 December 1993, 798–800.

80 On this point, see C. Bellamy, 'Big Five Nuclear Powers Braced for Move to Ban Bomb', *The Independent*, 4 January 1994.

81 International Court of Justice, *International Court of Justice Yearbook 1994–1995*, no. 49 (The Hague: International Court of Justice, 1995), 198.

82 United Nations, *Yearbook of the United Nations 1994* 48 (New York: United Nations Department of Public Information, 1995), 157–8.

Foreign Minister Evans expressed 'strong doubts about whether consideration by the ICJ of the question of the legality of nuclear weapons would in fact assist nuclear disarmament and non-proliferation'.[83]

Before 1995 the Labor government's position on nuclear disarmament could be described as equivocal. Although it reaffirmed Australia's commitment to the eventual elimination of nuclear weapons, the Keating government did not go any further than the Hawke government in elaborating a preferred timetable within which this objective might be attained. Notwithstanding the demise of the Cold War, and the subsequent re-emergence of nuclear disarmament as a popular topic of discussion, Australian policy makers maintained that incremental arms control measures deserved priority in the short to medium term over any serious consideration of nuclear disarmament.[84]

The shift in Australian nuclear policy during the second half of 1995 cannot be fully understood without reference to France's announcement that it would resume its nuclear testing program in the South Pacific. In what was later regarded as a major political miscalculation, the Keating government failed initially to comprehend the depth of indignation in the Australian electorate regarding the French announcement. Indeed, the source of community anger seemed to stem just as much from dissatisfaction with the Keating government's diplomatic response as from outrage at the French decision itself. The Keating government was placed on the political defensive from the outset.[85]

Aware of rising anger within the Australian electorate and among countries in the Asia-Pacific, the Keating government stepped up its condemnation in the week following the French announcement. Facing a national election in less than nine months, the government urgently sought to align its diplomatic posture more closely with prevailing community attitudes. Reasserting Labor's bona fides on arms control and disarmament was a central element of this repositioning.

In the most significant statement of Australian nuclear policy since Bill Hayden's *Uranium, the Joint Facilities, Disarmament and Peace* of 1983,[86] in October 1995 Keating delivered a speech to commemorate the fiftieth anniversary of

---

83 Editor, 'The ICJ and Nuclear Arms', *The Canberra Times*, 7 February 1995.

84 See for example Gareth Evans, 'Address to the 49th General Assembly of the United Nations by Senator the Hon. Gareth Evans QC, Minister for Foreign Affairs of Australia' (New York, 3 October 1994), 8–9, available at http://www.gevans.org/speeches2.html [accessed 2 July 2013].

85 For a thoughtful discussion of the domestic pressures on the government over the French testing decision, see Trevor Findlay, 'Explaining Australasian Angst: Australia, New Zealand and French Nuclear Testing', *Security Dialogue* 26, no. 4 (1995): 378–80.

86 Australian Department of Foreign Affairs, *Uranium, the Joint Facilities, Disarmament and Peace* (Canberra: Australian Government Publishing Service, 1984).

the United Nations. Reaffirming Australia's commitment to progress on nuclear arms control, Keating noted that the time had come 'for more determined movement towards the goal of a world free of nuclear weapons'.[87] As part of Labor's aim to 'advance the cause of a nuclear weapons-free world', Keating announced that the government would establish 'a group of knowledgeable and imaginative individuals from around the world' charged with formulating practical steps towards nuclear disarmament.[88]

While Keating's statement was generally well received, on the subject of nuclear disarmament the substance of the speech did not deviate from previous Australian governments' support for the *eventual* elimination of nuclear weapons. Keating's statement did not urge the nuclear powers to renounce the possession or use of nuclear weapons immediately. Nor did it recommend any preferred timetable for nuclear disarmament. Beyond announcing the Canberra Commission and a shift in Australia's position regarding the ICJ Case, there was little new in Keating's statement. Support for 'stable deterrence' pending the transition to disarmament echoed the view of the Hawke government that nuclear deterrence would pertain unless and until nuclear disarmament became a realistic policy option.

Nonetheless, nuclear disarmament was granted a historically unprecedented place in Australian foreign policy after October 1995. Never before had a government treated nuclear disarmament as being anything else other than strictly subordinate to progress in international nuclear arms control. Yet, a careful examination of Australia's subsequent involvement in the ICJ Case and the formation of the Canberra Commission reveal that the key substantive motives of the Keating government were still related to managing the risk of further nuclear proliferation.

Although the government's position on the ICJ Case announced by Keating appeared to amount to little more than a slight revision of Australia's position, it did represent a significant shift. While the Keating government would continue to argue that the ICJ should not make a decision on either request, if the Court eventually decided to formulate an advisory opinion, Australia would submit that it declare the use of nuclear weapons illegal. The government's willingness to countenance illegality as a 'fallback' position was directly at odds with the hard-line opposition to any ruling on the part of Australia's superpower ally, the United States.[89] In contrast to other Western countries such as Ireland, there

87 Paul Keating, 'The 50th Anniversary of the United Nations: Australia and a World Without Nuclear Weapons', Speech by the Prime Minister, the Hon. P.J. Keating, MP, Parliament House, Canberra (24 October 1995), 5.

88 Ibid., 8.

89 The significance of this was not lost on domestic supporters of illegality. See for example S. Wareham, 'Nuclear Weapons on Trial for the First Time in Their History', *The Canberra Times*, 30 October 1995.

was little public debate to speak of in Australia concerning the type of position the government should adopt at the ICJ.[90]

Thus the Keating government's position could not be attributed to alliance constraints or domestic pressures regarding the ICJ case specifically. Nor, given the qualifications contained in Keating's statement, could it readily be attributed to any newfound sympathy for the case for illegality. The shift owed more to recognition that the significance of the ICJ case went beyond international legal questions to encompass fundamental strategic issues about the role of nuclear weapons in international relations. By nominating illegality as Australia's fall-back position, the government sought to prevent a situation where certain non-nuclear states could point to a judgment of legality as further evidence of a global nuclear order in which the nuclear powers continued to dictate terms to the overwhelming majority of states who were expected to forego the acquisition of nuclear weapons in perpetuity.[91]

Less than a month after the government's oral submission to the ICJ, Keating announced the establishment of the Canberra Commission on the Elimination of Nuclear Weapons to be chaired by Australia's former Disarmament Ambassador, Richard Butler.[92] Though independent, the Commission would be supported by a secretariat based in the International Security Division of the Department of Foreign Affairs and Trade.[93] Couched in broad terms, the Commission's mandate required it to identify 'concrete and realistic steps for achieving a nuclear weapon free world, including the development and establishment of necessary verification and control mechanisms and new international legal obligations'.[94]

---

90 For discussion on this point see J. Carroll, 'Ireland Breaks EU Ranks and Supports Move to Ban Nuclear Weapons', *The Irish Times*, 10 June 1995; L. Kaucher, 'Nuclear Debate Needed', *The Sydney Morning Herald*, 15 September 1995.

91 'International Court of Justice, Requests for Advisory Opinions Submitted by the World Health Organization and the United Nations General Assembly, Oral Statements on Behalf of Australia, Opening Statement: The Hon. Gareth Evans QC, Minister for Foreign Affairs, 30 October 1995', 5, available at http://www.gevans.org/speeches/old/1995/301095_case_illegality.pdf [accessed 16 April 2015].

92 M. Grattan, 'PM Picks World Anti-Nuke Team', *The Age*, 27 November 1995.

93 Australian Department of Foreign Affairs and Trade, *Department of Foreign Affairs and Trade Annual Report: 1995–1996* (Canberra: Australian Government Publishing Service, 1996), 157.

94 Australian Department of Foreign Affairs and Trade, *Report of the Canberra Commission on the Elimination of Nuclear Weapons: August 1996* (Canberra: Australian Government Publishing Service, 1996), 107. An overview of the Commission Report presented to the Australian government on 14 August 1996 can be found in Andrew O'Neil, 'The Politics of Denuclearisation: Assessing the Canberra Commission Report', *Current Affairs Bulletin* 73, no. 2 (August/September 1996): 28–30.

Most observers chose to focus on the issue of nuclear disarmament as the sole yardstick by which the merits or otherwise of the creation of the Canberra Commission should be measured. Consequently, domestic criticism of the government's decision to form the Commission was more often than not based on the 'unrealistic' nature of the exercise, or the adverse impact Australian sponsorship would have on its alliance with the United States.[95] An evaluation of the various statements and, to a lesser extent, actions of the Keating government reveals that the formation of the Canberra Commission was not only due to conviction regarding the virtues of nuclear disarmament, but also due to concern over an increasingly precarious international environment for nuclear non-proliferation.

From the outset, the Keating government made it clear that it expected much of the Canberra Commission's analysis to focus on non-proliferation issues. Announcing the formation of the Commission in November 1995, Keating emphasized that a major part of the inquiry's scope would encompass 'political motivation and incentive in strengthening non-proliferation norms … the role of the nuclear weapons states and threshold states, [and] steps to strengthen international instruments and control mechanisms'.[96] Anxiety over nuclear proliferation trends in the event the nuclear powers failed to live up to their commitments to exercise restraint under the NPT was a major theme in Keating's remarks:

> Unless we take action [on nuclear disarmament] now and unless we address the issues of proliferation and the threshold states – that is those on the edge of getting nuclear weapons as well as nuclear terrorism – the nuclear competition which characterised much of the last half century will continue again after this pause. But one thing will be certain, it won't even have the stability of the bipolarity of the Cold War. It will be a multi-polar, unstable world where no-one is certain what the policy of deterrence really means and no-one [will] be able to guarantee what that action will really imply.[97]

Addressing the Commissioners at their first meeting in January 1996, Keating expressed concern regarding the possible 'emergence of a more fractured, dangerous world in which the restraints which have governed the strategic

95 See, respectively, D. Campbell, 'Utopian Gesture Will Not Disarm the Warheads', *The Australian*, 22 January 1996; and G. Sheridan, 'Disarmament Talk Endangers US Alliance', *The Australian*, 24 January 1996.

96 Paul Keating, 'Australian Initiative for a Nuclear Weapons-Free World: Statement by the Prime Minister, the Hon. P.J. Keating, MP, 26 November 1995', 2, available at http://pmtranscripts.dpmc.gov.au/transcripts/00009857.pdf [accessed 10 June 2014].

97 M. Bachelard, 'Introducing Paul Keating's First XV for the No Nukes Cup', *The Canberra Times*, 27 November 1995.

outlook since [nuclear] weapons were invented no longer apply, and the possession of such weapons proliferates more generally to other states and possibly even beyond state control'.[98]

From the Keating government's perspective, through its sponsorship of a Commission dealing explicitly with the topic of nuclear disarmament, Australia's reputation as an activist middle power could be enhanced. According to the government's logic, in the longer term the normative barriers to nuclear proliferation would only be maintained if the five nuclear powers displayed a genuine commitment to the principle of nuclear disarmament and exhibited a willingness to cut their respective nuclear arsenals. A renewed commitment to nuclear disarmament by the nuclear powers would help offset the sense of disenchantment among non-nuclear states while at the same time avoid compromising the perceived security benefits accrued by Australia as a result of its alliance with a nuclear-armed superpower. In that sense, the Canberra Commission and Australia's position in the ICJ case continued an Australian tradition of trying to bridge the gap between the NAM and nuclear powers in the NPT regime.

Conclusion

The end of the Cold War appeared to offer new hope for arms control and disarmament in the nuclear realm. For Australia, which had pursued a CTBT over many years, the first half of the 1990s presented an unprecedented opportunity to shape the multilateral agenda. Similarly, the terms and conditions for the discussion of nuclear disarmament entered a new phase in the 1990s. This seemed to promise a new era for substantial reductions in global nuclear stockpiles and a new discourse on the role of nuclear weapons in international relations.

However, for all the promise of this 'new world order', the Keating government remained fixated on the uncertainties and potential threats arising from the end of the Cold War. While the global picture had improved with the demise of the nuclear sword of Damocles enshrined in the 'central balance', Australia's region looked unsettled from a strategic perspective, and the Keating government had concerns about the prospects for nuclear proliferation in the Asia-Pacific. Indications of concern over the long-term credibility of US extended deterrence betrayed deeper misgivings about the strategic uncertainties in Australia's region. Such concerns were major factors in the Keating government's active global and regional diplomacy to see the fullest possible implementation of the IAEA's Additional Protocol and the indefinite extension of the NPT in 1995.

98 R. Gibson, 'Nuclear Experts Told to Act Now', *The Age*, 23 January 1996.

As this chapter shows, the Keating government's diplomatic activity on the CTBT and nuclear disarmament was in large part motivated by a concern that nuclear proliferation might accelerate after the Cold War in a context where the nuclear powers were unlikely to deliver on earlier pledges of nuclear restraint. While much of the government's rhetoric focused on internationalist goals in arms control and disarmament, the sources of Australia's nuclear policy during this period remained based on national security considerations. Never before had nuclear disarmament featured so prominently in Australian foreign policy, but it was intimately linked to strengthening the non-proliferation in a context where many non-nuclear states were increasingly vocal in their disenchantment with the behaviour of the nuclear powers.

Chapter 6
Great Power Politics, Proliferation and Terrorism: The Howard Government (1996–2007)

The 1996 national election ended 13 years of Labor Party rule in Australia, and returned the Liberal-National Coalition to power under Prime Minister John Howard. Howard had been Treasurer under Fraser. He was focused on trade and investment and had a close affinity to Australia's traditional allies, the United Kingdom and the United States. But Howard began his foreign policy with little experience or links – 'an amateur with attitude, passionate about his beliefs yet unsure of his policy', as Paul Kelly noted[1] – and it only became more central to the Howard government's agenda after the East Timor intervention of 1999. Howard's government continued to pursue the broad non-proliferation objectives that had defined Australian policy under the preceding Labor and Fraser governments. However, the Howard government placed far less emphasis than its predecessors on the promotion of nuclear disarmament as a 'practical' goal of Australia's foreign policy, which the Coalition believed had been an unnecessary point of contention in an otherwise smooth Australia-US relationship.[2]

The Howard government entered office as the initial optimism about the possibilities for arms control and global order in the post-Cold War era began to abate. In the nuclear sphere, Russia dragged out ratification of the START II treaty; India and Pakistan emerged as declared nuclear powers in 1998; actual and suspected nuclear programs in North Korea, Iraq, Iran and Syria challenged the success of the NPT regime, supported by proliferation networks emanating from North Korea and Pakistan's A.Q. Khan; the new administration of George W. Bush was far more willing than its predecessor to

1 Paul Kelly, *The March of Patriots: The Struggle for Modern Australia* (Melbourne: Melbourne University Press, 2009), 430.

2 For expressions of the new government's 'upgrading' of relations with the United States see Alexander Downer, 'Australia and the United States: A Vital Friendship', speech by Mr Alexander Downer, MP, Minister for Foreign Affairs, to the Australian Centre for American Studies, Sydney, 29 May 1996, available at http://www.dfat.gov.au/media/speeches/foreign/1996/aust_uni.html [accessed 10 July 2014].

reassess the value and role of nuclear weapons for US security in the post-Cold War world; and the Al-Qaeda attacks of 9/11 raised the spectre of nuclear terrorism. And all of these developments occurred in the context of a widening divide between NWS and NNWS members of the non-proliferation regime regarding the optimal balance between the aims of the NPT and a renewed interest, especially in Asia, in nuclear power for energy production. For some observers, these developments were harbingers of a 'second nuclear age', which would come to be characterized by the profound weakening of the institutions, practices and norms of behaviour established in each realm of nuclear policy during the Cold War.[3] The question for the Howard government during its time in office was whether, and how, the traditional 'grand bargain' approach to nuclear policy would have to adapt to the new global environment.

The Search for Strategic Stability in Post-Cold War Asia

Despite being initially lambasted by domestic critics that it did not 'get' Asia and was unduly wedded to an ANZUS-anchored vision of Australian security, the Howard government's emerging approach to the Asia-Pacific strategic environment exhibited broad continuity with that of the previous government.[4] The defining strategic fact for the incoming Howard government was that a post-Cold War order had yet to fully coalesce in the Asia-Pacific. Foreign Minister Downer, in one of his first major speeches, noted that while 'the Australian government does not see any direct security threat to Australia', much would depend on how relations between the regions' great powers evolved.[5] Downer also noted in this regard that these relationships could be shaped by the 'spill over' of long-standing regional disputes such as Taiwan,

---

3 See for example Jonathan Schell, 'The Second Age of Nuclear Danger', *Boston Review* (April/May 1998); Paul Bracken, 'The Second Nuclear Age', *Foreign Affairs* 79, no. 1 (2000); and for a more recent review of these issues see William Walker, *A Perpetual Menace: Nuclear Weapons and International Order* (London: Routledge, 2012), 124–33.

4 For critiques of the Howard government's initial years of foreign policy in Asia see for example Michael Wesley, *The Howard Paradox: Australian Diplomacy in Asia from 1996–2006* (Sydney: ABC Books, 2007), 7–8; Colin Brown, 'Problems in Australian Foreign Policy, January–June 1996', *Australian Journal of Politics and History* 42, no. 3 (1996): 331–44; and Mark Beeson, 'Australia and Asia: The Years of Living Aimlessly', *Southeast Asian Affairs* (2001): 33–44.

5 Alexander Downer, 'Australia and Asia: Taking the Long View', address by Mr Alexander Downer, MP, Minister for Foreign Affairs, to the Foreign Correspondents' Association, Sydney, 11 April 1996, available at http://www.foreignminister.gov.au/speeches/1996/asia-long.html [accessed 10 July 2014]. This also echoed the previous government's 1994 Defence White Paper assessment in many respects. See Australian Department of Defence, *Defending Australia: Defence White Paper 1994* (Canberra: Australian Government Publishing Service, 1994), 7–8.

the Korean Peninsula and the South China Sea, and new dynamics of WMD proliferation and international terrorism. Key to the development of strategic stability in post-Cold War Asia, Downer concluded, was the continued presence of the United States as it constituted 'the region's balancing wheel' and was 'overwhelmingly a positive force for regional stability' through its extensive security relationships and economic engagement throughout the region.[6] This demonstrated continuity with long-standing Australian security thinking that major strategic threats to the country's national security could 'arise as a consequence of distant disruption of the global balance of power' and that 'by choosing to work with more powerful allies to help ensure a satisfactory global balance, Australia thereby served its own interests'.[7]

The Howard government received both an early demonstration of the risks of this fluid strategic environment and an illustration of the continued importance of the US presence in Asia only days after it assumed office. Between March and April 1996, East Asia was plunged into one of the most serious diplomatic crises it had seen in decades as China conducted live-fire air and naval exercises in the Taiwan Straits between 8 and 25 March, prompting Washington to dispatch two carrier battle groups to the area. Significantly, Australia, alone among Washington's regional allies, publicly stated its unequivocal support for US actions in response to Beijing's 'belligerence',[8] plunging Sino-Australian relations into a 'decidedly frosty period'.[9] This crisis, so early in its tenure, underlined for the Howard government that rapid economic growth in Asia would present both an opportunity and challenges, as Cold War strategic moorings in the region were growing loose.[10]

---

6 Downer, 'Australia and Asia'.

7 Robyn Lim, 'Australian Security Thinking after the Cold War', *Orbis: A Journal of World Affairs* 42, no. 1 (Winter 1998): 95.

8 Foreign Minister Alexander Downer asserted at the time that: 'I think what we have seen in the last few days is a very clear demonstration by the United States that it is interested in maintaining its involvement in the security of the region and we obviously welcome that.' Quoted in Don Greenless and Richard MacGregor, 'Downer Warns China over War Games', *The Australian*, 13 March 1996.

9 The subsequent chill in Sino-Australian relations is covered extensively in (amongst many others) William Tow and Leisa Hay, 'Australia, the United States and a "China growing strong": Managing Conflict Avoidance', *Australian Journal of International Affairs* 55, no. 1 (2001): 37–54; and Roy C. McDowall, *Howard's Long March: The Strategic Depiction of China in Howard Government Policy, 1996–2006* (Canberra: ANU E-Press, 2009), 1–16.

10 Alexander Downer, 'Australia and Asia: Taking the Long View'; Alexander Downer, 'Security through Cooperation', address by Mr Alexander Downer, MP, Minister for Foreign Affairs, to the IISS/SDSC Conference: 'The New Security Agenda in the Asia-Pacific Region', Canberra, 2 May 1996, available at http://www.foreignminister.gov.au/1996/regsec5.html [accessed 10 July 2014].

The sense of regional instability fostered by the Taiwan Straits Crisis contributed to Washington and Canberra's reaffirmation of the relevance of the ANZUS alliance for Australian (and regional) security through the 'Sydney Declaration' at the inaugural Australia-US Ministerial Consultations in July 1996. The declaration reasserted both states' commitment to the obligations flowing from the ANZUS Treaty and identified the alliance's overarching goals of 'developing democracy, economic prosperity and strategic stability' in the Asia-Pacific.[11] Pointedly, 'strategic stability' was to be achieved by Australian and US efforts to prevent conflict, WMD proliferation and conventional arms build-ups in the Asia-Pacific.[12] Beijing, through its actions not only in the Taiwan Straits but also its June 1996 nuclear tests, and its opaque nuclear relationship with Pakistan, loomed large in this context,[13] although nuclear programs in North Korea and India also challenged the remnants of regional Cold War order.

Not surprisingly, and like its predecessor's 1994 Defence White Paper, the Howard government's 1997 *Strategic Review* thus confirmed that the ANZUS alliance did 'provide a clearer expectation of US support – that is, defence against nuclear attack'. Whilst the likelihood for a nuclear attack on Australia was judged to be 'very low', the review noted that in such a remote circumstance 'we would rely on the extended deterrence provided by the United States to deter such an attack.'[14]

Reference to nuclear disarmament, as either a goal or expectation, disappeared from Australian defence guidance documents. Instead, the 1997 *Strategic Review* was far more explicit at discussing the possibility of additional, future missile threats to Australia, stating that:

> One particular possibility which we will need to keep under review is the potential threat to Australia from ballistic and cruise missiles. Their long range and relative invulnerability to conventional air defences make them a potential

---

11 See Alexander Downer and Ian McLachlan, 'Australia-United States: A Strategic Partnership for the Twenty-First Century: Sydney Statement', Joint Security Declaration, Sydney, 27 July 1996, available at https://www.dfat.gov.au/geo/us/ausmin/sydney_statement.html [accessed 10 July 2014].

12 Ibid.

13 In the wake of the Sydney Declaration, for example, Beijing was to depict Australia as one of Washington's willing 'crab claws' (along with Japan) being used in a pincer movement to 'contain' China. McDowall, *Howard's Long March*, 14–16. For China's relations with Pakistan and nuclear proliferation concerns see for example Mohan Malik, 'Nuclear Proliferation in Asia: The China Factor', *Australian Journal of International Affairs* 53, no. 1 (1999): 31–41.

14 Australian Department of Defence, *Australia's Strategic Policy* (Canberra: Commonwealth of Australia, December 1997), 18.

> form of new threat to Australia. At present, Australia is within range of the ballistic missile forces of the five declared nuclear powers. And, in our region, India, Pakistan and North Korea have or are developing ballistic missiles. While none of those countries currently poses any threat to Australia, the proliferation of missiles within our region must be carefully watched.[15]

Strengthening alliance cooperation with the United States – including through cooperation on missile defence – was thus an essential factor for the Howard Government in Australia's post-Cold War security policy. The effect of this would, however, become far more apparent once the administration of George W. Bush defined its own particular approach to strategic and nuclear questions, than during Clinton's second term as US president from 1997 to 2001.

Strengthening the Non-Proliferation Regime: Additional Protocol and CTBT

Nonetheless, nuclear issues loomed large in the government's foreign policy from early on, with Foreign Minister Downer stating in September 1996 that 'nuclear proliferation and the attainment of nuclear weapons by rogue states ... may in time constitute *the greatest threat* to Australian and global security.'[16] As such, the Australian government would remain 'completely opposed to continued nuclear testing', 'fundamentally committed to the terms of the NPT', 'driving forward negotiations for the completion of the CTBT' and 'strongly oppose any action by existing nuclear weapons states which undermines the NPT and which would therefore weaken incentives for threshold countries to remain non-nuclear'.[17] For the Howard government, the maintenance of active non-proliferation diplomacy at the global level was intimately linked to the maintenance of a favourable regional environment as 'in addressing global problems, we are also helping to make our region secure'.[18]

Such rhetorical commitments were followed in practice with actions that underlined continuity rather than change in Australia's non-proliferation diplomacy. After China's June 1996 nuclear tests, Downer issued a statement condemning the tests and exhorting Beijing to 'desist immediately from all further testing' and 'move away from the now out-dated Cold War logic of nuclear confrontation'. Particularly irksome from Australia's perspective was the 'regrettable' timing of the tests as 'negotiations to conclude a CTBT are entering

---

15 Ibid., 31. The equivalent paragraph in the 1994 White Paper is on page 9 and only mentions missile proliferation without the reference to a direct threat to Australia.

16 Downer, 'Security through Cooperation'.

17 Ibid.

18 Ibid.

their final phase.' [19] In this respect the new government vigorously promoted the CTBT, sponsoring a resolution for the UNGA to endorse the text of the treaty in August 1996. As the Foreign Minster asserted, as Australia signed the treaty the following month, Australian activism had 'saved' the CTBT through its efforts in the UNGA. Moreover, Downer echoed the view of previous Australian governments that the successful conclusion of the CTBT would not only be a 'major milestone in global efforts to advance nuclear disarmament and non-proliferation objectives', but also be of 'particular significance for the countries of the Asia Pacific region – the only part of the world where all five nuclear weapons states have at some time conducted nuclear tests'.[20] In his address to the UNGA Downer also reaffirmed the four basic cornerstones of the country's traditional approach to nuclear proliferation and disarmament: prioritizing the NPT; support for the IAEA; bilateral safeguards agreements with other states party to the NPT; and an activist role in the multilateral non-proliferation forums.[21]

Even the government's presentation of the Canberra Commission to the Conference on Disarmament (CD), which was judged by most observers as unfavourable, in hindsight demonstrated continuity in substance with long-standing Australian policy, albeit with what would become the Howard government's typical tonal characteristics.[22] According to one critical view, Downer 'merely urged the CD to give the Report "careful consideration" and hoped that it would "stimulate international thinking"' on the disarmament agenda.[23] While the Howard government certainly remained more sceptical than its predecessor regarding the practicality of nuclear disarmament, especially in a fluid Asia-Pacific strategic environment, it nonetheless continued to vigorously advocate important 'practical' arms control measures and negotiations that were

---

19 See Alexander Downer, 'Chinese Nuclear Test', Department of Foreign Affairs and Trade Media Release, 8 June 1996, available at http://foreignminister.gov.au/releases/1996/fa46.html [accessed 1 June 2014].

20 Alexander Downer, 'Australia Signs the Comprehensive Test Ban Treaty (CTBT)', Department of Foreign Affairs and Trade Media Release, 25 September 1996, available at http://www.foreignminister.gov.au/1996/fa99.html [accessed 1 June 2014].

21 Alexander Downer, 'The United Nations for the 21st Century', address to the 51st General Assembly of the United Nations by Australia's Minister for Foreign Affairs, New York, 30 September 1996, available at http://www.dfat.gov.au/cc/947_3.html [accessed 30 June 2014].

22 For contemporaneous critiques see Richard Leaver, '"Middle Power Niche Diplomacy" and Nuclear Non-Proliferation: The Failure in Success', in Richard Leaver and Dave Cox, eds, *Middling, Meddling, Muddling: Issues in Australian Foreign Policy* (Sydney: Allen & Unwin, 1997), 182–201; and Hanson and Ungerer, 'The Canberra Commission: Paths Followed, Paths Ahead', 5–9.

23 Hanson and Ungerer, 'The Canberra Commission: Paths Followed, Paths Ahead', 10.

widely perceived to be crucial steps along the path towards nuclear disarmament. Thus in Downer's statement to the CD Australia also strongly urged the forum to focus on the arms control negotiations 'which are of most pressing concern to the international community' which in his view was the Fissile Material Cut-off Treaty.[24] Over the 1997 to 2000 period, the Howard government continued to advocate for the conclusion of this treaty, with Downer making annual statements in the CD and other multilateral forums to this end.[25]

Australia's early signing and ratification, as the first state to do so, of the IAEA's Additional Protocol in 1997 further demonstrated the Howard government's commitment to strengthen the non-proliferation regime in general, and in Southeast Asia in particular. The Additional Protocol had its genesis in revelations following the 1991 Gulf War that Iraq had managed to circumvent international safeguards in its quest to manufacture nuclear weapons. As noted in the previous chapter, the Keating government had been a major supporter of the '93+2' initiative and Australia played an important role in crafting the Additional Protocol and seeing its progress through the IAEA Board of Governors. In addition to extending safeguards to additional parts of the fuel cycle, the Additional Protocol endorsed a more detailed and intrusive set of verification measures regarding states' nuclear activities. This included states acceding to the Additional Protocol being required to provide the IAEA with greater levels of information through the 'Expanded Declaration' (encompassing information such as national nuclear fuel cycle development plans 10 years ahead), short-notice inspections of nuclear-related facilities.[26]

---

24 See Alexander Downer, 'Address by the Hon. Alexander Downer, MP, Minister for Foreign Affairs, to the Conference on Disarmament', Geneva, 30 January 1997, available at http://www.foreignminister.gov.au/speeches/1997/geneva_c.html [accessed 30 June 2014].

25 For example see Alexander Downer, 'Address by the Hon. Alexander Downer, MP, Minister for Foreign Affairs, to the Conference on Disarmament', Geneva, 30 January 1997, available at http://www.dfat.gov.au/media/speeches/foreign/1997/geneva_c.html [accessed 13 October 2006]; Alexander Downer, 'Two Swords for the Beating', address by the Hon. Alexander Downer, MP, Minister for Foreign Affairs, to the Conference on Disarmament, Geneva, 3 February 1998, available at http://www.dfat.gov.au/media/speeches/foreign/1998/swords3feb98.html [accessed 19 November 2006]; and Alexander Downer, 'The NPT – Future Challenges and Outlook for the 2000 Review Conference', keynote speech by the Hon. Alexander Downer, MP, Minister for Foreign Affairs, to the Conference on Nuclear Arms Control: Australian and Regional Perspectives Towards 2000, University of Queensland, 29 October 1999, available at http://www.dfat.gov.au/media/speeches/foreign/1999/991029_npt.html [accessed 27 November 2006].

26 See John Carlson, 'IAEA Safeguards Additional Protocol', 20 January 2009, available at https://www.dfat.gov.au/asno/publications/iaea-safeguards-additional-protocol.html [accessed 3 August 2014]; Richard Hooper, 'The IAEA's Additional Protocol', *Disarmament Forum* 3 (1999): 7–16.

At the regional level, Australia's endeavour to improve the rigour and reach of safeguards under the Additional Protocol was evident in diplomatic activity to encourage regional countries to sign up to the protocol and the Australian Safeguards Assistance Program (ASAP). The ASAP focused on training regional agencies to improve the safety and reliability of evolving national safeguards systems. Under the Howard government, ASNO developed outreach activities to help states prepare for changes to international safeguards standards resulting from the 93+2 initiative and the subsequent Additional Protocol. This included a course delivered to North Korean officials following the Howard's government's decision to normalize diplomatic relations with Pyongyang in 2000.[27] While these programs constituted an important element in Australia's nuclear policy, it should be emphasized that they were not well funded. Indeed, at the time, Australia's level of funding for safeguards outreach was less than 10% of what Canada spent on similar programs.[28]

In 2006, Australia gained the support of APEC ministers to explore the idea of regional safeguards cooperation to multilateralize what were a series of bilateral relationships. In 2009, this resulted in the creation of the Asia-Pacific Safeguards network comprising 14 safeguards-related organizations from 11 countries across the region.[29] Another key initiative by the Howard government, which complemented Australia's engagement in the Proliferation Security Initiative, was the 'Regional Security of Radioactive Sources' (RSRS) project launched in 2004. The RSRS project was coordinated by ANSTO in cooperation with the US Global Threat Reduction Initiative and the IAEA Office of Nuclear Security.[30]

A number of Foreign Minister Downer's statements in the CD and the UNGA in the late 1990s contained expressions of disappointment with the lack

27 See Australian Safeguards and Non-Proliferation Office, *Annual Report 2000–2001*, 'Output C: International Safeguards', available at https://www.dfat.gov.au/asno/annual_report_0001/output_c.html [accessed 1 October 2014].

28 David Anderson, 'Nuclear Safeguards', Parliamentary Research Service, Foreign Affairs, Defence and Trade Group, May 1997, available at http://www.aph.gov.au/Parliamentary_Business/Committees/Senate/Former_Committees/uranium/report/c12 [accessed 5 September 2014].

29 See John Carlson's paper 'Strengthening safeguards through regional cooperation: establishment of the Asia-Pacific Safeguards Network', presented at the Annual Meeting of the Institute for Nuclear Materials Management, Baltimore, July 2010, available at https://www.dfat.gov.au/asno/publications/strengthening-safeguards-through-regional-cooperation.html [accessed 1 October 2014].

30 The RSRS project was cancelled in 2013 following funding cuts by the Australian government. For background and analysis of the RSRS, see Tanya Ogilvie-White and David Santoro, 'Preventing Nuclear Terrorism: Australia's Leadership Role', *Australian Strategic Policy Institute Special Report* (Canberra: ASPI, 2014), 12–13.

of progress on some key facets of the international non-proliferation regime. These included the inability of the IAEA to adequately scrutinize and prevent the proliferation activities of states such as Iraq, Iran and North Korea, the weak international response to the nuclear weapons tests of India and Pakistan in 1998,[31] and the failure of the US Senate to ratify the CTBT in 1999.[32] The Howard government's response to the latter two developments in particular underscored Australia's enduring interest in limiting nuclear proliferation in its region. Australia immediately condemned India and Pakistan's tests and suspended official visits (including the recall of Australia's High Commissioners in each country), bilateral defence links and non-humanitarian aid.[33] Prime Minister Howard characterized the Indian tests as 'outrageous acts' and bemoaned New Delhi's desire for the 'grotesque status symbol' of nuclear weapons, while Downer argued that such tests 'could have most damaging consequences for security in South Asia and globally'.[34] Australia's Permanent Representative in the CD followed these reactions with a further statement of Australian opposition and condemnation of the nuclear tests. Australia's representative voiced the government's concern that the tests represented 'direct threats … to the credibility and existence' of the non-proliferation regime and that it hoped the international response would be 'swift and decisive' in order

31 For the weak international response see William Walker, 'International Nuclear Relations after the Indian and Pakistani Test Explosions', *International Affairs* 74, no. 3 (1998): 505–28.

32 See for example Alexander Downer, 'Address by the Hon. Alexander Downer, MP, Minister for Foreign Affairs, to the Conference on Disarmament', Geneva, 30 January 1997, http://www.dfat.gov.au/media/speeches/foreign/1997/geneva_c.html [accessed 10 August 2014]; Alexander Downer, 'Two Swords for the Beating', address by the Hon. Alexander Downer, MP, Minister for Foreign Affairs, to the Conference on Disarmament, Geneva, 3 February 1998, http://www.dfat.gov.au/media/speeches/foreign/1998/swords3feb98.html [accessed 10 August 2014]; and Alexander Downer, 'The NPT – Future Challenges and Outlook for the 2000 Review Conference', keynote speech by the Hon. Alexander Downer, MP, Minister for Foreign Affairs, to the Conference on Nuclear Arms Control: Australian and Regional Perspectives Towards 2000, University of Queensland, 29 October 1999, http://www.dfat.gov.au/media/speeches/foreign/1999/991029_npt.html [accessed 10 August 2014].

33 The Parliament of the Commonwealth of Australia, 'The 1998 Indian and Pakistani Nuclear Tests', Report of the Senate Foreign Affairs, Defence and Trade References Committee (Commonwealth of Australia, June 1999), 73–5, available at http://www.aph.gov.au/~/media/wopapub/senate/committee/fadt_ctte/completed_inquiries/1999_02/nuclear/report/report_pdf.ashx [accessed 10 August 2014].

34 See Press Release (Prime Minister), 'Indian Nuclear Tests', 12 May 1998; Press Release (Foreign Minister), 'Indian Nuclear Tests', 14 May 1998; Press Release (Foreign Minister) 'Pakistan's Nuclear Tests', 28 May 1998.

to 'send a message to other nations with possible nuclear aspirations about the consequences of such action'.[35]

For Australia, as a long-term champion of the CTBT, the failure of its ratification by the United States was also disheartening. Foreign Minister Downer, in his response to this failure, expressed the view of many close US allies and proponents of arms control by noting the difficulty of sanctioning the likes of India and Pakistan for nuclear testing while the United States failed to ratify the CTBT. Moreover, he strongly reasserted Australia's long-standing position as to the relevance of the CTBT, arguing that:

> The last thing the United States wants to see is a resumption of nuclear testing or the proliferation of nuclear weapons – and it is the last thing Australia wants to see. By refusing to ratify this treaty, the United States Senate has done a lot to undermine the arms control agenda that the international community, including Australia, has been working on.[36]

In his closing address to the 2000 NPT Review Conference, Foreign Minister Downer reinforced the perception that Australia not only remained an enthusiastic advocate of the regime, but that it did so in the belief that global action on non-proliferation was ultimately beneficial to Australian security:

> The NPT remains the world's best defence against the spread of nuclear weapons … As such it delivers major security benefits to all states, be they nuclear weapons states, non-nuclear weapons states, or even the four states yet to join the NPT.[37]

Clearly, the Howard government thus did not display the enthusiasm for nuclear disarmament that had been evident in the pronouncements of its Labor predecessors in general, and the Keating government's sponsorship of the Canberra Commission in particular. Nonetheless, it maintained strong support for the NPT regime, for the same reasons that had also made the Fraser government a vocal supporter: In the end, Australia's security strongly benefited from it.

---

35 Media Release, 'Australia's Statement to the Conference on Disarmament', 2 June 1998, available at http://www.dfat.gov.au/media/releases/department/d055_98.html [accessed 10 August 2014].

36 Quoted in 'Aftermath of US Senate Rejection of CTBT', *Disarmament Diplomacy* 41 (November 1999), available at http://www.acronym.org.uk/dd/dd41/41after.htm [accessed 10 August 2014].

37 See Alexander Downer, 'Speech by the Minister for Foreign Affairs', NPT Review Conference, New York, 25 April 2000, available at http://www.dfat.gov.au/media/speeches/foreign/2000/000425_npt.html [accessed 10 August 2014].

Confronting Global Terror, 'Rogue States' and Proliferation Networks

The broad question that confronted the Howard government at the turn of the century, however, was whether the broader nuclear order as it had survived from the days of the Cold War required adapting for the new post-Cold War era. Specifically, would a nuclear order based on states' deliberate commitment to nuclear deterrence and nuclear abstinence be enough to provide strategic stability and avoid the use of nuclear weapons?[38] Two developments, the election in 2000 of a new Republican administration under George W. Bush that was determined to create a national missile defence program sooner rather than later, and the spectre of nuclear terrorism after the 9/11 attacks, both raised this issue in their own way. After 2001, both would lead to a significant shift in the US, and hence also the Australian, approach to managing the risk of nuclear proliferation.

The election of George W. Bush brought a Republican Party to power in the United States that had largely been convinced for a number of years of the need to field a National Missile Defence (NMD) capability. In 1995, the Republican-dominated Congress had passed legislation mandating the deployment of a limited NMD by 2003,[39] which set the scene for a continued tussle with a reluctant Clinton administration throughout the remainder of the decade. The 1998 report of the bipartisan 'Commission to Assess the Ballistic Missile Threat to the US' (or the 'Rumsfeld Commission' after its chairman, Donald Rumsfeld),[40] reinforced arguments for the urgency of deploying a limited shield against 'rogue state' ICBM. With the election of George W. Bush, and the return to the Pentagon of Donald Rumsfeld as Secretary of Defense, it seemed only a matter of time before the United States would abrogate the ABM treaty and field a national missile defence system. Speaking in May 2001, President Bush declared that 'Today's world requires a new policy, a broad strategy of active proliferation, counter-proliferation and defenses', and that the United States required:

> a new framework that allows us to build missile defenses to counter the different threats of today's world. To do so, we must move beyond the constraints of the 30 year old ABM treaty. This treaty does not recognize the present, or point us to the future. It enshrines the past. No treaty that prevents us from addressing

---

38 Walker, *A Perpetual Menace*, 130–33.

39 See Lewis, Gronlund and Wright, 'National Missile Defense: An Indefensible System', *Foreign Policy* (Winter 1999/2000): 121; Michael O'Hanlon, 'Star Wars Strikes Back', *Foreign Affairs* 78, no. 6 (1999): 77–8.

40 Michael O'Hanlon, *Defense Policy Choices for the Bush Administration*, 2nd edition (Washington, DC: Brookings Institution Press, 2002), 127. For an overview of the commission's report see The Commission to Assess the Ballistic Missile Threat to the United States, *Executive Summary of the Commission to Assess the Ballistic Missile Threat to the United States* (Washington, DC: U.S. Government Printing Office, 15 July 1998), available at http://fas.org/irp/threat/bm-threat.htm [accessed 15 July 2014].

> today's threat that prohibits us from pursuing promising technology to defend ourselves, our friends and our allies is in our interest or in the interest of world peace. This new framework must encourage still further cuts in nuclear weapons. Nuclear weapons still have a vital role to play in our security and that of our allies.[41]

Conscious of the perceptions that this shift in US policy caused both internationally and within Australia, the Howard government balanced its reaction between general support to the United States on the one hand, and avoiding an explicit endorsement of the Bush administration's plans on the other. In March 2001, Foreign Minister Downer argued during debate in federal parliament that Australian support for the United States in this matter was essential to maintain the health of the ANZUS alliance.[42] Defence Minister Peter Reith also asserted in a June 2001 address that Australia had adopted a 'position of understanding US plans to develop a limited missile defence system to defend against the potential threats posed by the proliferation of missile technology' and reaffirmed 'our enduring role, extending back nearly as far as ANZUS – some thirty years – in the early warning business'. Cooperation in BMD-related research and development would continue to provide Australia with access to data through the Joint Facilities; it would have potential for US BMD systems to protect ADF personnel deployed overseas; and it could generate other technological benefits and applications for the ADF.[43]

In December 2001, the United States formally withdrew the ABM Treaty, and President Bush directed the Department of Defense to accelerate the testing and deployment of NMD capabilities in January 2002.[44] After the 9/11 attacks, the administration's evolving approach to nuclear (and other WMD) proliferation placed significant emphasis on the sovereign rights and responsibilities of states to deal with proliferation, be it by state or non-state actors,[45] and the Howard government broadly accepted this approach.

The clearest indication of this was in the government's public statements prior to the 'coalition of the willing' invasion of Iraq in March 2003.

41 The White House Office of the Press Secretary, 'Remarks by the President to Students and Faculty at National Defense University', Fort Lesley J. McNair, Washington, DC, 1 May 2001, available at http://www.whitehouse.gov/news/releases/2001/05/20010501.html [accessed 10 August 2014].

42 *C.P.D.* (Representatives), 8 March 2001, 25483.

43 See Peter Reith, 'Address by the Hon. Peter Reith, MP, Minister for Defence to the ANZUS Conference', 30 June 2001, available at http://www.defence.gov.au/minister/8tpl.cfm?CurrentId=785 [accessed 10 August 2014].

44 Bernd W. Kubbig, 'America: Escaping the Legacy of the ABM Treaty', *Contemporary Security Policy* 26, no. 3 (2005): 424.

45 See Jofi Joseph, 'The Exercise of National Sovereignty: The Bush Administration's Approach to Combating Weapons of Mass Destruction Proliferation', *The Nonproliferation Review* 12, no. 2 (2005): 374.

Australia's commitment to this endeavour was justified on the basis of the government's belief that Iraq's WMD capabilities posed an immediate threat to international security, either through Iraq's possible use of them or their transfer into hands of a terrorist organization.[46] However, it also carried the implication that the Australian government also subscribed to the Bush administration's policy of pre-emptive use of force against a state believed to have both contravened its obligations under the NPT and to embody the nexus between the greatest threats to international security (i.e. rogue states, WMD proliferation and terrorism). Howard clearly set out this position in his address to the National Press Club on 14 March 2003, stating that:

> We believe that it is very much in the national interest of Australia that Iraq have taken from her chemical and biological weapons and denied the possibility of ever having nuclear weapons. Not only is it inherently dangerous for a country such as Iraq with its appalling track record to have these weapons but if Iraq is allowed to get away with it other rogue states will believe they can do the same … And as these dangerous weapons spread so the risk that they may fall into the hands of terrorists will multiply. And if terrorists ever get their hands on weapons of mass destruction that will, in my very passionate belief and argument, constitute a direct, undeniable and lethal threat to Australia and its people, and that would be the ultimate nightmare not only for us but for other peoples in other nations. That, more than anything else, is the reason why we have taken the stance we have and it's the reason why we believe that Iraq should be effectively and comprehensively disarmed.[47]

Moreover, this address was imbued with a profound scepticism of the ability of the NPT and its UN-sanctioned multilateral mechanisms to not only discipline Iraq but also other states suspected of violating their non-proliferation responsibilities under the NPT such as North Korea and Iran.[48] Subsequent government statements and actions indicated further that Australia was re-evaluating long-held policy positions, such as the priority given to the NPT, in relation to nuclear proliferation. Foreign Minister Downer, for example, while maintaining that the non-proliferation regime remained the 'mainstay for stopping the spread of WMD', simultaneously argued that

46 See Australian Broadcasting Corporation (ABC Online), available at http://www.abc.net.au/news/2014-09-21/howard-embarrassed-by-intelligence-that-iraq-had-wmds/5759132 [accessed 1 October 2014].

47 John Howard, 'Transcript of the Prime Minister the Hon. John Howard MP Address to the National Press Club', The Great Hall Parliament House, Canberra, 14 March 2003, available at http://www.pm.gov.au/news/speeches/speech/74.html [accessed 11 November 2006].

48 Ibid. As noted above these concerns had also been expressed for some time prior to 9/11.

the international community 'should be open to new methods and new thinking' as the cause of non-proliferation was 'too important to be bound by rigid dogma'.[49]

Such 'new thinking' was best embodied in President George W. Bush's 'Proliferation Security Initiative' (PSI) of 2003. The initiative was formed in response to heightened concerns after 9/11 regarding illicit trade in WMD and ballistic missile-related materials and technology. This threat was underlined by an incident in 2002, where a North Korean freighter, the *So San*, was found by a Spanish Navy boarding team to contain Scud missiles bound for Yemen,[50] and the unravelling of the A.Q. Khan nuclear proliferation network in 2004 after centrifuge parts bound for Libya were found on a freighter in Italy.[51] Both of these cases highlighted the need for measures to prevent trafficking of nuclear materials that evaded export controls, through better international and domestic coordination and information exchange, better understanding of the legal instruments available to inspect international shipping and the closing of loopholes in international law against WMD proliferation (especially through binding UN Security Council resolutions).

Australia was a strong and original supporter of the initiative, which President Bush had privately discussed with Howard before its public launch.[52] Although PSI arose from the same set of considerations that framed the Bush administration's decision to invade Iraq, namely the potential nexus between WMD proliferation, 'rogue' states and international terrorism,[53] it was far less controversial and countries that had opposed the invasion of Iraq, including Germany and France, were also original members. This was largely due to the

---

49 Alexander Downer, 'Weapons of Mass Destruction: The Greatest Threat to International Security', speech by the Hon. Alexander Downer, MP, Minister for Foreign Affairs, to the Proliferation Security Initiative, Brisbane, 9 July 2003, available at http://www.foreignminister.gov.au/speeches/2003/030709_wmd.html [accessed 10 August 2014].

50 Andrew C. Winner, 'The Proliferation Security Initiative: The New Face of Interdiction', *The Washington Quarterly* 28, no. 2 (2005): 131; Michael Byers, 'Policing the High Seas: The Proliferation Security Initiative', *The American Journal of International Law* 98, no. 3 (2004): 526–8.

51 For detailed analysis of the A.Q. Khan Network see Braun and Chyba, 'Proliferation Rings: New Challenges to the Nuclear Non-Proliferation Regime'; and Adrian Levy and Catherine Scott-Clark, *Deception: Pakistan, the United States and the Global Nuclear Conspiracy* (London: Atlantic Books, 2007).

52 David Cooper, 'Challenging Contemporary Notions of Middle Power Influence: Implications of the Proliferation Security Initiative for "Middle Power Theory"', *Foreign Policy Analysis* 7, no. 3 (2011): 324.

53 Ron Huisken, 'The Proliferation Security Initiative: Coming in from the Cold', *APSNet Policy Forum*, 20 April 2006, available at http://nautilus.org/apsnet/0613a-huisken-html/ [accessed 10 June 2013].

fact that PSI was not designed to replace existing non-proliferation regimes, but to enhance and enforce them. Australia's own involvement was motivated by three major factors: its shared desire to counter further ballistic and WMD proliferation, especially in its immediate region; its role as a close US ally in the Asia-Pacific; and its large maritime jurisdiction.[54] Australia's hosted the first PSI exercise, Pacific Protector 03, in which participating countries practiced intercepting, boarding and searching suspicious vessels.[55] The second meeting of PSI members took place in Brisbane in July 2003.[56] It addressed practical issues such as information sharing, operations and training exercises, but participating nations also recognized that in order to be effective the PSI required broad membership, particularly by states in which cargo ships tend to be registered, coastal states and transit states.[57] This was especially the case in archipelagic Southeast Asia where Australia's participation and diplomatic support could help assuage political and legal concerns about the initiative,[58] and Australia took a regional leadership role.[59]

---

54 See Emma Belcher, 'A Tighter Net: Strengthening the Proliferation Security Initiative', *Lowy Institute Policy Brief*, August 2009, 4, available at http://www.lowyinstitute.org/publications/tighter-net-strengthening-proliferation-security-initiative [accessed 10 June 2013]; and Clive Schofield et al., 'Securing Maritime Australia: Developments in Maritime Surveillance and Security', *Ocean Developments and International Law* 39, no. 1 (2008): 95.

55 See Australian Department of Defence website, available at http://www.defence.gov.au/psi/ [accessed 2 January 2013]. In addition, it also hosted subsequent exercises Pacific Protector 06 and Pacific Protector 10.

56 Mark J. Valencia, 'The Proliferation Security Initiative: Making Waves in Asia', *The Adelphi Papers* 45, no. 376 (2005): 26.

57 National Institute for Public Policy, 'The Proliferation Security Initiative: A Model for Future International Collaboration', *Comparative Strategy* (2009), 411.

58 Susan Koch, 'Proliferation Security Initiative: Origins and Evolution', *Center for the Study of Weapons of Mass Destruction Occasional Paper No. 9* (Washington, DC: National Defense University Press, 2012), 8, available at http://www.ndu.edu/press/lib/pdf/CSWMD-OccasionalPapers/CSWMD_OccationalPaper-9.pdf [accessed 9 September 2013]. When the PSI was first formed, its membership was composed almost exclusively of Western nations, with Japan the only exception. Australia's diplomatic support in early years included sending Paul Sullivan, Deputy Secretary of Foreign Affairs, to Beijing to explain the details of the PSI to the Chinese government. In addition, Canberra expended a great deal of effort in convincing Japan to participate in the first PSI exercise, Pacific Protector 03, addressing Tokyo's sensitivities regarding the military nature of the exercise. Australian officials also attended the 2003 Asian Senior-Level Talks in Non-Proliferation and, in concert with the US and Japanese delegations, delivered a detailed explanation of the PSI. Newman and Williams, 'The Proliferation Security Initiative', 307, 308, 314.

59 Cooper, 'Challenging Contemporary Notions of Middle Power Influence: Implications of the Proliferation Security Initiative for "Middle Power Theory"', 328.

The Howard government continued to embrace the Bush administration's strategy of seeking alternative non-proliferation mechanisms beyond those of the NPT during its remaining time in office. Most Australian security agencies began to acknowledge a counterproliferation role, with DFAT in a coordinating role. In 2004, Foreign Minister Downer announced that Australia would focus its counterproliferation efforts in three areas, which mirrored those of the Bush Administration: the Proliferation Security Initiative, ballistic missile defence and 'practical measures to strengthen the global counter-proliferation architecture'. In justifying this, Downer argued that the international community had 'to go beyond the constraints imposed by consensus decision making' in order to ensure 'compliance and enforcement in these regimes'.[60] The Foreign Minister's statements at the 2005 NPT Review Conference also further underlined the government's waning commitment to Australia's 'traditional', consensus-oriented approach towards the NPT. Most notably Australia's long-standing position that the onus was on *both* the NNWS and NWS to uphold the NPT and further its goals had shifted to a position where, in Downer's words:

> We acknowledge progress in reducing nuclear arsenals, but expect further steps by the nuclear weapons states. But we do not accept that movement on nuclear disarmament should be a precondition for improvements to the non-proliferation regime. Such an approach puts at risk the security benefit all NPT parties derive from assurances that nuclear programs in non-nuclear weapons states are peaceful.[61]

This clearly aligned Australia with the immediate imperatives of the United States and its allies in preventing Iran (an NPT party) and North Korea (until 2003 an NPT party), from utilizing Article IV of the NPT to pursue the acquisition of nuclear breakout capabilities, let alone nuclear weapons. The notion that the NWS should not have to show further progress on the question of disarmament as a *quid quo pro* for the NNWS adherence to non-proliferation standards did reduce Australian pressure on the United States to sign the CTBT. At the same time, however, it also stood in a long tradition of Australian nuclear policy that placed particular emphasis on the practical implementation and success of NPT commitments – beginning with the Fraser government's support for strong safeguards.

---

60 See Alexander Downer, 'The Threat of Proliferation: Global Resolve and Australian Action', speech to the Lowy Institute by the Hon. Alexander Downer, MP, Minister for Foreign Affairs, Sydney, 23 February 2004, available at http://www.foreignminister.gov.au/speeches/2004/04223_lowy.html [accessed 20 March 2007].

61 Alexander Downer, 'Statement to the Seventh Nuclear Non-Proliferation Treaty Review Conference by the Australian Minister for Foreign Affairs, the Hon. Alexander Downer MP', New York, 2 May 2005, available at http://www.foreignminister.gov.au/speeches/2005/050502_7th_nuclear_conference.html [accessed 30 March 2007].

Supplying Nuclear Fuel to Asia's Great Powers

In the final two years in office of the Howard government, a clear break with the policy settings of Australia's nuclear grand bargain was implemented in relation to the export of uranium to India. At the core of Australia's bargain had been the assumption that strategic and diplomatic benefits would accrue to Australia through a 'principled' uranium export policy, which required recipients (other than the five recognized nuclear powers) to become member states of the NPT and accept IAEA as well as Australian bilateral safeguards. The Howard government's response to the conclusion of the George W. Bush administration's nuclear cooperation agreement with India in March 2006 clearly challenged this orthodoxy. Whereas Australia's role as a provider of energy security to recipient countries had previously been confined within the NPT context, in the case of India the strategic and economic benefits from becoming a supplier of uranium to one of Asia's rising great powers – and potential counterweight to China – reversed this order of policy priorities.

After the finalization of the US-India agreement in New Delhi, Foreign Minister Downer asserted that Australia welcomed it and believed that it was a 'good agreement', as it would draw 'India into the mainstream of the international community'.[62] Prime Minister Howard concurred suggesting that the agreement was 'a very positive development'.[63] In stark contrast to his strong condemnation of the Indian nuclear tests in 1998, Howard simply noted that 'you never say never' when asked whether Australia would reconsider its own uranium export position.[64] At a joint press conference with visiting US Secretary of State, Condoleezza Rice, Downer stated that India giving up its nuclear weapons and signing the NPT as NNWS was 'a dream' that would not 'happen any time soon … regardless how idealistic and passionate we may be about that', and voiced Australian support for the 'logic of what the United States administration has done' with respect to India.[65]

---

62 Alexander Downer, 'Transcript of the Foreign Minister the Hon. Alexander Downer MP, Doorstop Interview', Hyatt Hotel, Adelaide, 3 March 2006, available at http://www.foreignminister.gov.au/transcripts/2006/060303_ds.html [accessed 15 June 2007].

63 John Howard, 'Transcript of the Prime Minister the Hon. John Howard MP, Doorstop Interview', Taj Mahal Hotel, Delhi, 5 March 2006, available at http://www.pm.gov.au/news/interviews/Interview1806.html [accessed 5 December 2006].

64 Geoffrey Barker, 'PM on Uranium: Never Say Never', *The Weekend Australian Financial Review*, 18–19 March 2006.

65 Alexander Downer, 'Joint Press Conference between Foreign Minister Alexander Downer and US Secretary of State, Condoleezza Rice', 16 March 2006, available at http://www.foreignminister.gov.au/transcripts/2006/060316_joint_press_conference.html [accessed 15 June 2007].

While Downer did not explain what this logic was, there were two logics – relating to non-proliferation and strategic considerations – underpinning the US-India deal. The first stemmed from the Bush Administration's perception that one of the chief problems of the NPT regime was its discriminatory nature. The fact that the regime permitted an NPT party such as Iran to engage in proliferation activities while enjoying the right to import nuclear materials and technologies for 'peaceful purposes', while denying the same right to a non-party such as India that was deemed to have a 'good' proliferation record, underlined the shortcomings of the non-proliferation regime in aligning incentives with behaviour. The strategic logic of the agreement was also clearly expressed in the major statements of the administration officials in relation to the agreement. Despite the peppering of President Bush and Secretary of State Rice's comments, for example, with references to the 'shared' democratic values of the US and India, the agreement's effective acceptance of India as a legitimate nuclear power was arguably geared towards affecting the regional and international strategic environment.[66] Thus, the subsequent talk of a 'strategic partnership' between Washington and New Delhi, encompassing not only the nuclear agreement but also increased military, security and economic relations, suggested that the Administration was pursuing a wider strategic agenda to maintain a balance between the rising powers of China and India, and the United States.[67]

Less than a month after Howard and Downer's comments on the US-India agreement, Australia concluded two agreements with China regarding the transfer of nuclear material and nuclear cooperation. These agreements committed Australia and China to 'cooperate in the peaceful uses of nuclear energy' including 'collaboration in the transfer of nuclear material' and the 'transfer of material, equipment, components, and technology, research and development, exchange of information and projects of mutual interest'.[68] Export of nuclear materials (i.e. uranium) to China was consistent with Australia's long-standing uranium export policy as China was both a recognized NWS under the

66 The White House Office of the Press Secretary, 'President, Prime Minister Singh Discuss Growing Strategic Partnership', 2 March 2006, available at http://georgewbushwhitehouse.archives.gov/news/releases/2006/03/images/20060302-9_p030206pm-0310-515h.html [accessed 12 January 2011].

67 See for example Ashley Tellis, 'The Evolution of US-Indian Ties: Missile Defense in an Emerging Strategic Relationship', *International Security* 30, no. 4 (2006): 113–51; Ashton B. Carter, 'America's New Strategic Partner?', *Foreign Affairs* 85, no. 4 (2006): 33–44.

68 Australian Department of Foreign Affairs and Trade, *Agreement between the Government of Australia and the Government of the People's Republic of China for Cooperation in the Peaceful Uses of Nuclear Energy*, Canberra, 3 April 2006, available at http://www.dfat.gov.au/geo/china/treaties/nuclear_energy.html [accessed 10 August 2014].

NPT and had ratified the Additional Protocol.[69] Despite this, however, there remained questions regarding the timing of this agreement.[70] In particular, if China satisfied Australia's requirements regarding safeguards of nuclear material why had not such a deal been concluded prior to 2006? Two major considerations arguably entered the Howard government's calculations in this context. First, China had only entered the non-proliferation mainstream in the 1990s, acceding to the NPT in 1992, signing the CTBT in 1996 and signing and then ratifying the Additional Protocol in 1998 and 2002 respectively.[71] A second major point of consideration concerned the broader dynamics of the ANZUS alliance in the context of a rising China, with which Australia had developed a relationship based upon (from Australia's perspective) a clear calibration of mutual economic interests. Indeed, Prime Minister Howard often argued in the 2000s that Australia could develop increasingly close relations without jeopardizing Australia's security and strategic relationship with Washington.

In this context, the conclusion of the Australia-China agreement of 2005 was not only an economic opportunity for Australia to supply the growing Chinese nuclear power program, but also an opportunity to reduce the possible irritant value, in the context of Sino-Australian relations, of Australia's association with the US approach to non-proliferation and counterproliferation. China's immediate response to the PSI, for example, had been rather ambivalent, voicing agreement with the PSI's objectives and principles,[72] while also stating that 'China holds that the fundamental purpose of non-proliferation is to safeguard and promote international and regional peace and security, and all means to this end should be conducive to attaining this goal,'[73] which meant that 'Unilateralism and double-standards must be abandoned.'[74] In addition, China was sensitive in terms of its own proliferation record and activities,[75]

69 Australian Department of Foreign Affairs and Trade, 'Australia-China Nuclear Material Transfer Agreement and Nuclear Cooperation Agreement, FAQ', available at http://www.dfat.gov.au/geo/china/treaties/faq.html [accessed 10 August 2014].

70 John Garnaut, 'China Breaches Spur Nuclear Fallout', *The Age*, 27 August 2007, available at http://www.theage.com.au/news/business/chinese-breaches-spur-nuclear-fallout/2007/08/26/1188066943726.html [accessed 10 August 2014].

71 Information Office of the State Council, *China's Endeavors for Arms Control, Disarmament and Non-Proliferation*, March 2005, available at http://english.xinhua.cn [accessed 10 August 2014]; see also Wendy Frieman, *China, Arms Control and Non-Proliferation* (London: Routledge, 2004).

72 Newman and Williams, 'The Proliferation Security Initiative', 311.

73 Information Office of the State Council of the People's Republic of China, *China's Non-Proliferation Policy and Measures* (Beijing: New Star Publishers, 2003), 1–2.

74 Ibid., 3–4.

75 For example see Frieman, *China, Arms Control and Non-Proliferation*, 5–7; Alastair Iain Johnston, 'Learning Versus Adaptation: Explaining Change in Chinese Arms Control Policy in the 1980s and 1990s', *The China Journal* 35 (January 1996): 30.

including sales of nuclear materials or technologies to Iran and Pakistan and allegations of Chinese nuclear espionage in the United States.[76] In this context, the agreement allowed Prime Minister Howard to state his support for China's credentials, saying that he was 'satisfied that the safeguards that are there will be enforced and it's on that basis that the agreements have been signed'.[77]

Following the signing of bilateral safeguards agreements with China, the Howard government then became more explicit in its support for the approach underpinning the US-India agreement, with Downer stating that:

> This whole India issue, it's a very difficult question because you can have the status quo – or becoming the status quo ante now – where India is not an NPT party, it has nuclear programs, it has nuclear weapons and you can take the view therefore we will have nothing to do with them. Or you can take the view that the Bush Administration has taken that, well, at least you can embrace some of India's nuclear industry and you can have inspections by the IAEA of some of those nuclear facilities. And some inspections and some transparency is better than none, isn't it?[78]

Additionally, it was now clear that the Howard government viewed the potential of exporting uranium to India under such conditions as not only beneficial with respect to non-proliferation concerns but also as a means to secure a strategic relationship with New Delhi. Indeed, Prime Minister Howard, when asked if Australia would consider selling uranium to India along the lines of the US deal in March 2007, noted that 'We see India as a very responsible country. The relationship between Australia and India is growing. It is a very important relationship. They will be the considerations that we will bear in mind.'[79] On 15 August 2007, only weeks before the election, the government then decided in principle to commence negotiations about a bilateral agreement governing the sale of uranium to India.[80] While there were non-proliferation arguments

---

76 See Frieman, *China, Arms Control and Non-Proliferation*, 20; Mohan Malik, 'Nuclear Proliferation in Asia: The China Factor', *Australian Journal of International Affairs* 55, no. 1 (1999): 31–41.

77 See Howard, 'Transcript of the Prime Minister, the Hon. John Howard, MP, Joint Press Conference with His Excellency Mr Wen Jiabao'.

78 Alexander Downer, 'Speech and Question and Answer Session, Energy Environment and Air Quality Policy Forum', Los Angeles, 13 January 2007, available at http://www.foreignminister.gov.au/transcripts/2007/070113_eeaqpf.html [accessed 13 September 2007].

79 Katherine Murphy, 'Howard's U-Turn on India', *The Age*, 30 March 2007, available at http://www.theage.com.au/articles/2007/03/29/1174761667423.html [accessed 5 June 2007].

80 See for example Michael Clarke, 'Australia, India and the Uranium Question', *Australian Journal of Political Science* 46, no. 3 (2011): 487–500; Sandy Gordon,

both for and against this decision, it is clear that the strategic argument about India's importance for Asian geopolitics – and hence also for Australia itself – weighed strongly in this decision which was the first clear break of an Australian government with Fraser's original uranium export policy.

Reconsidering the Economic Domain: Towards the UMPNER Review

One of the enduring concerns for Australian governments since Fraser was the maintenance of Australia's nuclear technical, scientific and regulatory expertise as a prerequisite for the operation of the grand bargain. This was particularly well demonstrated by deliberations across the Keating and Howard governments regarding the potential replacement of ANSTO's HIFAR research reactor and the general state of Australia's nuclear science and technology base.

In August 1993 the Keating government had received the findings of the Report of the Research Reactor Review (RRR), or 'McKinnon Review', which had been charged with reviewing the costs and benefits for 'commercial, scientific and national interests reasons' of a new research reactor for Australia. The RRR, despite noting specific scientific and technical limitations of HIFAR due to its age, recommended that 'there was no pressing need to make a decision on the fate of the reactor' and that the reactor's 'remaining life was likely to extend beyond a decade'.[81] The Keating government subsequently accepted the RRR's recommendations in November 1993 and decided to delay a final decision on replacement of HIFAR 'for about five years'. Simultaneously however it also committed to the establishment of a 'Nuclear Expertise Steering Group' to 'oversee a new program of training and development in nuclear fuel cycle matters to improve the availability of expertise in these areas within the Government'.[82] Foreign Minister Gareth Evans maintained that such an initiative was required for Australia 'to be able to fully keep pace with technological developments, understand what is happening elsewhere and make an appropriate contribution to the international policy debate as a result'.[83] Despite this, however, the level of funding provided for such a program remained low with an allocation of just $115,000 in 1995.[84]

When the Howard government came into office, it therefore inherited the responsibility for the 'final decision' regarding the replacement or otherwise of

---

'Implications of the Sale of Australian Uranium to India', Working Paper No. 410 (Canberra: Strategic and Defence Studies Centre, September 2008).

81 K.R. McKinnon et al., *Future Reaction: Report of the Research Reactor Review* (Sydney: Australian Government Publishing Service, 1993), xiii.

82 Media Release, Senator Chris Schacht, Minister for Science and Small Business, Minister for Assisting the Prime Minister for Science, 9 November 1993.

83 Foreign Affairs, Defence and Trade Legislation Committee, 1 June 1995, 18/23.

84 Ibid.

HIFAR as well as concerns regarding the health of Australia's nuclear science and technology base. Its decision of September 1997 to replace HIFAR thus rested upon consideration of not only the scientific and research benefits to Australia of a new research reactor but also the broad 'national interest' implications of the retention of significant nuclear science expertise for Australia's capacity to contribute to global non-proliferation efforts.[85] DFAT provided a clear statement of such considerations to the Senate Economics References Committee in May 1998 when it noted that the operation of a research reactor 'underpins Australian technical nuclear expertise and that in turn is an essential input into policy advice to the government across a range of nuclear issues'. More pointedly, DFAT identified that operation of a research reactor enabled Australia 'to play a significant international role' through its seat on the Board of Governors of the IAEA and was essential for Australia's ability to contribute to the strength of nuclear safeguards and evaluations of safeguards compliance.[86]

In February 1999 the House of Representatives referred the issue to the Parliamentary Committee on Public Works for review, which concluded that:

> There is a continuing need to ensure that the immediate environment remains free of nuclear weapons. As one of the world's major exporters of uranium, there is a need to maintain a leading role in the international safeguards system. For these national interest requirements to have credibility, the Government must have technical expertise to pursue policies on nuclear issues and non-proliferation. ANSTO and the Department of Foreign Affairs and Trade advised the committee that Australia's credibility as a regional leader in nuclear issues is becoming difficult to maintain, citing the lack of a modern research reactor as the prime reason.[87]

---

85 Chapter 2, 'HIFAR and the Replacement Research Reactor Project', available at http://www.aph.gov.au/~/media/wopapub/senate/committee/lucasheights_ctte/report/c02_doc.ashx [accessed 20 July 2013].

86 Remarks by Ms Deborah Anne Stokes, Department of Foreign Affairs and Trade, to the Economics References Committee, 'Reactor at Lucas Heights', 11 May 1998, Senate Hansard, E146, available at http://www.aph.gov.au/binaries/hansard/senate/commttee/s1746.pdf [accessed 20 July 2013]. The DFAT representative explicitly noted that without the technical knowledge acquired from the operation of a research reactor Australia would have been unable to assist the IAEA in the strengthening of safeguards after the discovery of the Iraqi nuclear program in the wake of the First Gulf War or be in a position to evaluate proliferation threats from states such as North Korea.

87 Commonwealth of Australia, 'Proposed Replacement Nuclear Research Reactor, Lucas Heights, NSW', *Parliamentary Standing Committee on Public Works* (Canberra: Commonwealth of Australia, 12 August 1999), 16.

In addition, if Australia failed to maintain its scientific and technical expertise (including through the replacement of HIFAR), the DFAT advice continued, 'the basis for Australia's retention of its seat on the Board of Governors of the IAEA would be undermined,' negatively impacting on the country's 'ability to influence developments in international nuclear affairs, including nuclear safeguards'.[88] The Committee subsequently judged HIFAR to be obsolete and concurred with advice provided by DFAT, ASNO and ANSTO that a new reactor would meet the 'national interest considerations' detailed above.[89] The Howard government proceeded with plans to construct a new reactor and on 6 June 2000 announced that the Argentine company INVAP S.E. had been chosen as the successful tenderer for what would become the Open-pool Light Water Reactor (OPAL),[90] which became operational in 2007.

The Howard government was thus clearly not averse to investment into nuclear technology. However, despite OPAL's role in the supply of medical isotopes, it remains first and foremost a scientific and research facility, and although ANSTO remains involved in national security research – including the support of safeguards through the Safeguards Support Program and IAEA reference laboratory – its role in economic development remains significantly curtailed compared to its predecessor, the Australian Atomic Energy Commission.

During the 2000s, however, the economic and non-proliferation arguments about an increasing Australian involvement the fuel cycle itself once again gained significant salience. A major factor related to changes in the international nuclear energy market that affected Australia's position as a supplier of uranium. The burgeoning economic growth of China (and also India) and the concomitant increase in demand for energy, combined with growing international concern about the effects of global warming and access to, and security of, existing energy resources, contributed to a reconsideration of nuclear energy internationally. Many states, particularly in Asia, developed plans to either expand existing civilian nuclear reactor fleets or establish civilian nuclear power programs.[91] Australia, possessing the world's largest recoverable

88 Ibid.

89 Ibid., 21.

90 Ibid., 22.

91 See for example John Deutch, Arnold Canter, Ernest Moniz and Daniel Poneman, 'Making the World Safe for Nuclear Energy', *Survival* 46 (Winter 2004/05): 66; Massachusetts Institute of Technology, *The Future of Nuclear Power: An Interdisciplinary MIT Study* (Cambridge, MA: Massachusetts Institute of Technology, 2003), available at http://web.mit.edu/nuclearpower/ [accessed 10 June 2013]; Chaim Braun, 'The Nuclear Energy Market and the Nonproliferation Regime', *The Nonproliferation Review* 13, no. 3 (2006): 627–44; Trevor Findlay, 'The Nuclear Energy Revival and Global Governance', in Michael Clarke, Stephan Frühling and Andrew O'Neil, eds, *Australia's Uranium Trade: The Domestic and Foreign Policy Challenges of a Contentious Export* (Farnham: Ashgate, 2011), 15–38.

deposits of uranium, thus stood to gain substantial export earnings if such rising demand for uranium occurred. Indeed, the Howard government's 2004 White Paper *Securing Australia's Energy Future* reflected emerging enthusiasm for the potential of Australia's uranium resources in this international context by noting that the development of Australia's 'low cost energy resources', including uranium, would be 'key to our future'.[92]

A second, related factor was the launch by the US government of its Global Nuclear Energy Partnership (GNEP) initiative in February 2006. GNEP in its initial formulation envisaged to facilitate nuclear power adoption and at the same time strengthen non-proliferation safeguards through the provision of nuclear fuel in exchange for countries' renunciation of sovereign enrichment or reprocessing capabilities.[93] From Australia's perspective, GNEP (much as the INFCE had in the late 1970s) constituted both an opportunity and a challenge. Potentially, GNEP could provide Australia with a major opportunity to establish itself as a reliable fuel supply partner for developing countries, but alternatively it could lead to Australia's exclusion from the list of countries engaging in enrichment, and hence prevent it from downstream 'value-adding' to its uranium reserves.[94] Liberal Senator Russell Trood voiced this concern during Senate debate when he noted that while GNEP held the potential to limit the expansion of uranium enrichment capabilities, a non-proliferation goal broadly supported by Australia, Australia 'should be careful about this' as:

> It is not obvious that that proposal necessarily includes Australia, because, at the moment, we do not enrich uranium, but if it were to pass and we were not part of it then our own capacity to enrich and add value to our own uranium reserves would be affected.[95]

It was in this context that Prime Minister Howard asserted in July 2006 that Australia needed to consider pursuing such 'value adding' to its uranium. This had followed the Howard government's earlier decision to support the commercial enrichment activities of the Australian company SILEX Systems under strict

92 Australian Department of Prime Minister and Cabinet, *Securing Australia's Energy Future* (Canberra: Commonwealth of Australia, 2004), 3.

93 Jeff Lindemeyer, 'The Global Nuclear Energy Partnership: Proliferation Concerns and Implications', *The Nonproliferation Review* 16, no. 1 (2009): 80.

94 Andrew Davies, 'Australian Uranium Exports and Security: Preventing Proliferation', *ASPI Strategic Insight* (August 2006): 14–15. For similar concerns in the 1970s see Wayne Reynolds, 'Australia's Quest to Enrich Uranium and the Whitlam Government's Loans Affair', *Australian Journal of Politics and History* 54, no. 4 (2008): 562–78; Wayne Reynolds, 'The Yellow Cake Road: Malcolm Fraser, the Ranger Enquiry and Australia's Role in the US International Fuel Cycle Project', *Australian Journal of Politics and History* 57, no. 4 (2011): 511–25.

95 *C.P.D.* (Senate), 'Matters of Public Interest – Nuclear Non-proliferation', 21 June 2006, 56.

regulations in partnership with the United States. To accommodate the joint initiative, Washington and Canberra concluded a revision to the 1981 bilateral nuclear cooperation agreement.[96] To safeguard the technological development, which included the transfer of uranium hexafluoride – and presumably to minimize publicity of the sensitive nature of the laser enrichment – the United States and Australia sought to restrict access to the activities. This led to accusations from Greenpeace that the Howard government was contributing to the risk of proliferation by endorsing activities that allegedly had a weapons application.[97] By 2006, due to a lack of funding in Australia, the SILEX technology was sold to the US corporation General Electric.[98]

For Prime Minister John Howard, Australia with its substantial uranium resources had the potential to become an 'energy superpower' if it implemented the 'right policies' and a 'head in the sand' approach to its engagement in the nuclear fuel cycle in particular would be 'akin to Saudi Arabia turning her back on global oil developments'.[99] Yet Foreign Minister Downer, careful to note that the government was supportive of the Bush administration's preference to limit the number of states with uranium enrichment capabilities, suggested that 'an approach worth considering is to build a regime that allows for enrichment based on a range of strict criteria' including:

> judgements on a country's track record and whether it is in a sensitive and dangerous region and therefore enrichment would have a destabilising effect … Our Government will make sure that any decision on enrichment would be internationally responsible and in line with preserving regional and international security, indeed enhancing it. We certainly wouldn't want this to become a cause for concern in the region or amongst our allies.[100]

---

96 Minister for Foreign Affairs Alexander Downer, 'SILEX: Australia/US Nuclear Cooperation', *Media Release*, 26 May 2000.

97 In particular, see *Greenpeace, Secrets, Lies and Uranium Enrichment: The Classified SILEX Project at Lucas Heights* (Sydney: Greenpeace, November 2004), available at: http://www.mapw.org.au/download/secrets-lies-and-uranium-enrichment-2004-greenpeace-report-silex [accessed 1 October 2014].

98 Silex, *Annual Report 2013*, 14, available at: http://www.silex.com.au/downloads/asxannouncements/17-slx-annual-report-2013-fa-online [accessed 1 October 2014].

99 See for example Mischa Schubert, 'PMs Vision of Australia as Energy Superpower', *The Age*, 18 July 2006, available at http://www.theage.com.au/news/business/pms-vision-of-australia-as-energy-superpower/2006/07/17/1152988472470.html [accessed 5 July 2014]; Raymond Bonner, 'Call to Enrich Uranium Stirs Debate in Australia', *New York Times*, 2 August 2006, available at http://www.nytimes.com/2006/08/02/world/asia/02australia.html?pagewanted=print&_r=0 [accessed 5 July 2014].

100 Alexander Downer, 'The Impact of Energy Security on Australia's International Relations', speech to the Australian Institute of International Affairs, 11 October 2006, available at http://www.foreignminister.gov.au/speeches/2006/061011_es.html [accessed 5 July 2014].

These considerations prompted the government to undertake a review into Australia's engagement in the nuclear fuel cycle, the *Uranium Mining, Processing, and Nuclear Energy Review* (UMPNER), under the chairmanship of Dr Ziggy Switkowski. UMPNER's final report, delivered in December 2006, canvassed the potential expansion of Australia's exports of yellowcake and the requirements needed for the establishment of civil nuclear power and uranium enrichment industries in Australia. With respect to enrichment it has since been reported that European enrichment consortium URENCO discussed details of a potential partnership arrangement with Australia in which URENCO would construct and manage an enrichment facility in Australia at the cost of approximately $2 billion.[101] The UMPNER report also suggested that Australia could play a role in receiving and storing radioactive waste products.[102]

The UMPNER report noted that the potential expansion of Australia's engagement in the nuclear fuel cycle carried a number of downsides due to four major unresolved problems of nuclear energy: costs, safety, proliferation and waste.[103] The cost and safety factors were primarily related to the construction, maintenance and secure operation of civil nuclear power plants and are not considered to be a principal risk in relation to security and proliferation concerns.[104] Public and governmental concerns regarding the effect of increased Australian uranium exports, enrichment of uranium and development of a civil

---

101 Jeffrey Lantis, 'Elections and Enduring Realities: Australia's Nuclear Debate', *Arms Control Today* (April 2008), available at https://www.armscontrol.org/act/2008_04/Lantis [accessed 5 July 2014].

102 See Australian Department of Prime Minister and Cabinet, *Uranium Mining, Processing and Nuclear Energy – Opportunities for Australia?* (Canberra: Commonwealth of Australia, 2006), 2–3, 101–8. However, the historical record suggested that the government would have a difficult task to persuade the Australian public of the benefits of developing such an industry. Indeed, the Howard government had faced significant controversy and widespread public opposition in 1998 regarding the proposed development by Pangaea Resources of an international, high-level radioactive waste storage and disposal facility in a remote region of Western or South Australia. Ultimately the Howard government had in this particular instance adopted the position that there was in fact 'no relationship between benefiting from selling uranium and taking responsibility for the waste produced by its use'. Ian Holland, 'Waste Not Want Not? Australia and the Politics of High-level Nuclear Waste', *Australian Journal of Political Science* 37, no. 2 (2002): 283–301.

103 See Australian Department of Prime Minister and Cabinet, *Uranium Mining, Processing and Nuclear Energy – Opportunities for Australia?*, 3–7; *The Future of Nuclear Power: An Interdisciplinary MIT Study*, 2; Jacques Percebois, 'The Peaceful Uses of Nuclear Energy: Technologies of the Front and Back-Ends of the Fuel Cycle', *Energy Policy* 31 (2003): 101–8.

104 John Deutch, Arnold Canter, Ernest Moniz and Daniel Poneman, 'Making the World Safe for Nuclear Energy', 66.

nuclear energy industry, however, stemmed from the greater proliferation risks associated with the increased presence of nuclear materials these steps would entail.[105] While such concerns ultimately underpinned much of the public and media reaction to the release of the UMPNER report, Prime Minister Howard remained convinced of the need for Australia to expand not only its engagement in the nuclear fuel cycle but also look closely at the development of a civil nuclear industry.[106] To that end, the Howard government approved ANSTO's development of a collaborative nuclear research program that was aimed in part at enhancing knowledge within Australia about new 'Generation IV' nuclear reactor technology and promoting closer cooperation between ANSTO and Australian universities.[107]

However, these were not views shared by the new Labor government that came to power less than a year after the release of the UMPNER report, and the new Prime Minister Kevin Rudd committed to repealing the agreement with India and shelving any consideration of uranium enrichment or civil nuclear power development.

Conclusion

When the Howard government came into office in March 1996 there was little indication that it would diverge from Australia's long-standing policy-settings on nuclear issues, which were built on the bedrock assumption that Australia could use its uranium export policy as an instrument to achieve its three core interests in this sphere: to prevent further horizontal proliferation in the Asia-Pacific; to reap the commercial benefits of its uranium resources while abiding by its commitments under the NPT; and to enhance its reputation as an activist middle power. By the early 2000s, however, the Howard government began to reconsider some of the core pillars of the grand bargain, including the priority accorded to the NPT in Australian diplomacy and whether Australian exports should be linked to it in the special case of India. These changes were, in the main, stimulated by the rise of China and India as great powers; continued nuclear proliferation in Asia and beyond, despite the near-universality of the NPT; and the advent of a new, globally oriented Islamic terrorism.

105 Kate O'Neill, 'Radioactive "Trade": Globalizing the Nuclear Fuel Cycle', *SAIS Review*, no. 1 (Winter-Spring 2002): 157–68.

106 For example Katherine Murphy, 'Howard's Backyard Nuclear Blitz Opens Election Bidding', *The Age*, 30 December 2006, available at http://www.theage.com.au/news/national/howards-nuclear-backyard-blitz-opens-election-bidding/2006/12/29/1166895479310.html [accessed 10 August 2013].

107 See ANSTO, 'ANSTO Collaborative Nuclear Research Program', 2007, available at http://www.ansto.gov.au/__data/assets/pdf_file/0018/15921/NPP_QandA.pdf [accessed 1 October 2014].

As a result, Howard was willing to place more emphasis on outcomes than established norms in his government's approach to nuclear policy – a stance that united his approach to the pre-emptive use of force in Iraq, the establishment of the PSI as an enduring 'coalition of the willing' rather than a formal, multilateral organization, and the recognition of India's status as a de-facto nuclear weapons state outside the NPT. In all of these aspects, the government followed and supported US policy, but Australia's traditional geostrategic priorities, especially regarding the robustness of the non-proliferation regime in Southeast Asia, remained a continuing concern for his government as well. As such, the main characteristic of the Howard government's approach, especially after the turn of the century, was a deliberate willingness to modernize and adapt Australia's nuclear policy to what it perceived as the new, enduring realities of the 21st-century international environment.

Chapter 7

Consensus and Stagnation in Australian Nuclear Policy: The Rudd-Gillard-Rudd Governments (2007–13)

On 24 November 2007, the Labor Party, led by Kevin Rudd, defeated the Howard government at the national election. Rudd came to office with a strong connection to younger sections of the electorate and a declared commitment to significant policy shifts. One of the first actions of the new government was to endorse and ratify the Kyoto Protocol and confirm Australia's commitment to further measures to address climate change. Yet, less than three years later, in a stunning political coup, Rudd would be out of office. Rudd was ejected by his senior ministerial colleague Julia Gillard who emerged in June 2010 to become Australia's first female Prime Minister. Facing disastrous public opinion polls in June 2013, Gillard in turn was replaced as Prime Minister by Kevin Rudd in a party ballot, and he subsequently led Labor to electoral defeat the same year.

The approach to foreign and strategic policy by the newly elected Rudd government was based on three key pillars: deepening Australia's alliance with the United States, greater engagement in the Asia-Pacific region and strengthening Australian integration in the UN system. Overlaying these three pillars would be a return to Labor's traditional emphasis on multilateralism and international institutions as the primary means by which Australia pursued its global interests. The tone and substance of the Rudd government's foreign policy was thus closer to the Keating government than to the Hawke government. Whereas the latter had stressed continuity with the previous Fraser government, the Rudd government was keen to signal a clear break with its Coalition predecessor, and references to 'Labor tradition' in foreign policy were frequent under the new government.

Nuclear policy was no exception to this pattern, and the purpose of the International Commission on Nuclear Non-proliferation and Disarmament (ICNND) created by the Rudd government was justified partly in terms of completing what Keating's Canberra Commission had started. The new government rejected the recommendation of the UMPNER review that Australia should begin developing the basic elements of a domestic nuclear power industry, and one of its early decisions was to disestablish the body charged with implementing the key recommendations of the 2006 report.[1] Likewise, the Rudd government came to office committed to overturning Howard's decision to export uranium to India.

One of the characteristics of the Rudd government in general was the combination of policy centralization with limited attention to implementation. Hence, its nuclear policy exhibited the same dysfunctionalities and short-termism apparent in other fields.[2] Andrew Carr writes that 'other than the ICNND, the Rudd government made no significant investment of resources into the cause, and as the prime minister grew steadily more distracted by domestic events and his rating in opinion polls dropped during 2009 and especially 2010, non-proliferation and non-possession efforts seemed to be ignored or abandoned as a government priority.'[3] When Rudd was reportedly invited by the White House to speak at a Nuclear Security Summit in 2010, he chose not to given domestic policy priorities.[4]

Andrew Carr notes that in relation to non-proliferation, Rudd and Gillard 'were of the same views, though whereas Rudd was highly activist, it was a virtual non-issue for Gillard'.[5] The only main point of difference in both governments' nuclear policy – and at the same time the only main nuclear policy initiative of the Gillard government – was to overturn Rudd's refusal to proceed with the Howard era decision to allow export of uranium to India. In their own way, therefore, both Rudd and Gillard's nuclear policies were backward rather than forward looking: Rudd's to 'Labor activism' in general and the Keating government in particular; Gillard's to overturning Rudd's policy on India. As a result, Labor's tumultuous six years in office represented the culmination, but also stagnation of more than half a century of Australian nuclear policy.

A Mirage of Nuclear Disarmament Policy

The first months of the Rudd government were dominated by the public launching of a host of reviews, initiatives and domestic summits in a whole range of areas, and nuclear policy was no exception. In June 2008, Prime Minister

1 Masakatsu Oda, 'Japan Lobbied for more Nuclear Weapons', *Kyodo News*, 23 November 2009.

2 See Andrew Carr and Chris Roberts, 'Foreign Policy', in Chris Aulich and Mark Evans, eds, *The Rudd Government: Australian Commonwealth Administration, 2007–2010* (Canberra: ANU E-Press, 2010).

3 Andrew Carr, *Winning the Peace: Australia's Campaign to Change the Asia Pacific* (Melbourne: Melbourne University Press, 2014), 188.

4 Daniel Flitton, 'Abbott's "Anglosphere" Looks to the Past', *Sydney Morning Herald*, 27 April 2010, available at www.smh.com.au/federal-politics/political-opinion/abbotts-anglosphere-looks-to-the-past-20100426-tn7l.html [accessed 6 August 2014].

5 Carr, *Winning the Peace*, 185.

Rudd announced that Australia would co-sponsor with Japan the creation of the International Commission on Nuclear Non-proliferation and Disarmament (ICNND). The role of the ICNND would be to 'pick up' the work of the Canberra Commission on the Elimination of Nuclear Weapons.[6] As noted in the previous chapter, the Howard government had accepted the report of the Canberra Commission unenthusiastically. Co-chaired by former Australian Foreign Minister Gareth Evans and former Japanese Foreign Minister Yoriko Kawaguchi, the ICNND's role would be to 'reinvigorate, at a high political level, awareness of the global need for nuclear non-proliferation and disarmament', with a view to strengthening the prospects for a positive outcome at the 2010 NPT Review Conference and promoting nuclear restraint in the longer term.[7]

It is notable that the ICNND initiative was launched the year before President Obama's Prague speech, during the final year of the Bush administration. Advocacy of disarmament was at the time increasing in many quarters around the globe, but the Bush administration had shown little genuine interest in de-emphasizing the role of nuclear weapons in American strategic policy. In fact, between 2001 and 2009, the administration sought to substantially reinvest in US nuclear capabilities after what many Republicans had described as a period of neglect under the Clinton administration. While continuing significant reductions in the level of strategic weapons systems in accordance with the arms control agreements with Russia, the Bush administration made it clear to allies and adversaries alike that Washington regarded the nuclear umbrella as an enduring feature of alliance commitments.

President Barack Obama's landmark speech in Prague in April 2009 that affirmed America's commitment to nuclear disarmament signalled what appeared to be a major shift in US policy, and made the creation of the ICNND particularly prescient.[8] US governments had made rhetorical commitments to nuclear disarmament in the past, but President Obama's speech coincided with unprecedented enthusiasm among US and European elites for deep cuts to nuclear weapons on the road to eventually eliminating all nuclear weapons. High profile interventions by former senior US policy officials in favour of disarmament were replicated in the UK and Germany, although the remaining

---

6 Australian Broadcasting Corporation (ABC Online), 'Rudd Heralds New Nuclear Disarmament Body', *ABC News Online*, 9 June 2008, available at http://www.abc.net.au/news/2008-06-09/rudd-heralds-new-nuclear-disarmament-body/2465316 [accessed 10 July 2010].

7 'International Commission on Nuclear Non-proliferation and Disarmament: Joint Statement by Gareth Evans and Yoriko Kawaguchi, New York, 25 September 2008', available at http://www.mofa.go.jp/ICSFiles/afieldfile/2008/09/26/h2009_nichi_2.pdf [accessed 10 July 2010].

8 The White House, 'Remarks by President Obama', Hradcany Square, Prague, 5 April 2009, available at http://www.whitehouse.gov/the_press_office/Remarks-By-President-Barack-Obama-In-Prague-As-Delivered [accessed 09 September 2014].

nuclear powers including France, China and Russia remained conspicuously silent on the issue.[9]

President Obama's Prague speech was publicly welcomed by US allies, but privately there were misgivings about the potential implications for US nuclear doctrine and the nuclear umbrella in the lead up to the 2010 US Nuclear Posture Review (NPR). There was a lively debate in Washington about whether the US should embrace a no-first-use doctrine as part of the NPR, which was given impetus by the need for a positive outcome at the 2010 NPT review conference following an unsuccessful conference outcome in 2005. Misgivings were especially acute in Japan where policy makers lobbied Washington intensively throughout the US Nuclear Posture Review process.[10] Japanese lobbying for the Obama administration *not* to embrace a no-first-use commitment was especially intense in a context where Tokyo's relations with Beijing were deteriorating and North Korea's nuclear program was accelerating. Tokyo was especially keen for the United States to send a message to Pyongyang and Beijing that nuclear weapons might be reserved for use in the event of conventional aggression. And Japanese policy makers were very focused on the need to avoid any perception that Washington's resolve to use nuclear force to support allies might be weakening.[11] This was in line with a long-standing trend where 'even barely perceptible signs of weakness in the US nuclear posture have had the potential to trigger alarm and overreaction in Tokyo'.[12]

The Rudd government's endorsement of a major international project to promote nuclear disarmament thus sat somewhat uncomfortably with the increasing strategic uncertainty in Asia. Of all America's allies in Asia, Australia has historically been the least anxious about potential shifts in US nuclear policy and doctrine, but the Rudd and Gillard governments nevertheless watched closely as the Obama administration formulated the 2010 NPR. It is telling therefore that whereas the creation of ICNND had been ahead of Obama's push for disarmament, the Rudd Government did little to maintain the momentum and follow up opportunities created by the ICNND.

---

9 The most salient contribution to this thinking was a 2007 article in *The Wall Street Journal* by four former senior American policy makers. See George Schultz, William Perry, Henry Kissinger and Sam Nunn, 'A World Free of Nuclear Weapons', *The Wall Street Journal*, 4 January 2007.

10 See Masa Takubo, 'The Role of Nuclear Weapons: Japan, the U.S., and "Sole-Purpose"', *Arms Control Today*, November 2009, available at https://www.armscontrol.org/act/2009_11/Takubo#7 [accessed 09 September 2014].

11 Oda, 'Japan Lobbied for more Nuclear Weapons'.

12 Toshi Yoshihara and James Holmes, 'Thinking about the Unthinkable: Tokyo's Nuclear Option', *Naval War College Review* 62, no. 3 (2009): 65.

The final ICNND report was released in December 2009 as *Eliminating Nuclear Threats: A Practical Agenda for Global Policymakers.*[13] Its tone was more focused on policy realities than the 1996 Canberra Commission Report, and it contained a considerable amount of technical detail. Indeed, its central recommendations that the global nuclear stockpile be reduced to 2,000 warheads by 2020 and that all nuclear powers embrace a no-first-use doctrine were hardly radical measures given the ultimate objective of disarmament it recommended. Australia and Japan developed from the report a joint 'Package of Practical Nuclear Disarmament and Non-Proliferation Measures' for consideration at the 2010 NPT Review Conference. However, the timing of the Australian government's response to the ICNND report demonstrated a desire to be in sync – or at least not out of step – with American nuclear policy. It took until the NPT Review Conference in May 2010 for the Australian government to release an official response to the report that was long in praise, but did little to maintain any kind of political momentum.[14]

In contrast, the Foreign and Defence Minister released a joint response to the 2010 US Nuclear Posture Review only a day after it had been released in April 2010. Whereas the release stated that 'much of the [ICNND] report's disarmament agenda is consistent with government policy', it was notably more effusive in its praise of the US NPR, calling it 'a comprehensive, forward looking and pragmatic approach to nuclear non-proliferation and disarmament, tailored to meet the challenges and threats of the 21st century'.[15] This was despite the fact that the NPR did not take up the ICNND recommendation of a comprehensive no-first-use assurance. Nor did the lack of detail over quantitative nuclear reductions in the Nuclear Posture Review replicate the comparatively detailed vision outlined in the ICNND.

In the end, the Rudd government shared this lack of enthusiasm for detailed commitment and timelines for nuclear weapons reductions, which was an approach continued by the Gillard government. Indeed, there is evidence the Gillard government was even more ambivalent about nuclear disarmament than its predecessor. In April 2013, at an intergovernmental-NGO conference in

---

13 International Commission on Nuclear Non-proliferation and Disarmament, *Eliminating Nuclear Threats: A Practical Agenda for Global Policymakers* (Canberra: Paragon, 2009).

14 See 'Australian Government's response to the International Commission on Nuclear Non-proliferation and Disarmament (ICNND) Report', Media Release, Stephen Smith, MP, Minister for Foreign Affairs and Trade, 3 May 2010, available at http://www.foreignminister.gov.au/releases/2010/fa-s100503a.html [accessed 15 September 2014].

15 See 'The United States' Nuclear Posture Review', Joint Media Release, Stephen Smith, MP, Minister for Foreign Affairs, and Senator John Faulkner, Minister for Defence, 7 April 2010, available at http://foreignminister.gov.au/releases/2010/fa-s100407.html [accessed 15 September 2014].

Norway, Australian officials actively (though unsuccessfully) sought to block any reference in the final document to a Red Cross submission that characterized any use of nuclear weapons as causing 'incalculable human suffering'. Gareth Evans subsequently branded the Gillard government's approach 'completely indefensible',[16] but documents released under Freedom of Information laws showed that senior Australian officials believed that endorsing such a statement would be incompatible with continuing support for US extended nuclear deterrence.[17] Moreover, in October 2013 Australia chose not to support a statement on the humanitarian consequences of nuclear weapons put forward at the UN General Assembly by New Zealand Ambassador Dell Higgie, on behalf of 123 countries and the Holy See. Australia argued that the statement gave insufficient recognition to the security dimensions of nuclear weapons. Australia instead released its own joint statement on behalf of 17 countries (of which most are recipients of extended nuclear deterrence), arguing the need to engage 'substantively and constructively' with nuclear weapons states with recognition to 'both the security and humanitarian dimensions of the nuclear weapons debate'.[18]

Extended Deterrence in a Contested Asia

Despite embracing many of the liberal internationalist themes of previous Labor governments, including in relation to nuclear disarmament, the Rudd government exhibited a much sharper realist edge when it came to strategic policy, especially in relation to potential shifts in Asia's geopolitical landscape. The rise of China's influence in Asia throughout the 2000s had taken place at the same time as Washington had been preoccupied with two wars in the Middle East. Australian strategists remained concerned about perception of American decline, which had gained traction in parts of the region as China began to invest heavily in a major military modernization program and demonstrated growing confidence in asserting its claims over maritime territory in Southeast and Northeast Asia. This apprehension shone through in the 2009 Defence

16 Quoted in Philip Dorling, 'ALP Nuclear Backflip Linked to US Defence', *The Sydney Morning Herald*, 2 October 2013.

17 Ibid.

18 Ramesh Thakur, 'The Differing No-Nuke Stance between Australia and New Zealand', *Worldviews*, 29 November 2013, available at http://passblue.com/2013/11/29/the-differing-no-nuke-stance-between-australia-and-new-zealand/ [accessed 12 November 2014]. *Joint Statement on the Humanitarian Consequences of Nuclear Weapons, Delivered by Ambassador Peter Woolcott, Australian Permanent Representative to the United Nations, Geneva and Ambassador for Disarmament*, UNGA 68 First Committee, 21 October 2013, available at http://www.un.org/disarmament/special/meetings/firstcommittee/68/pdfs/TD_21-Oct_CL-1_Australia-(Joint%20St).pdf [accessed 12 November 2014].

White Paper, which Rudd himself was closely involved in drafting. It included unusually direct language that referred to China's lack of transparency potentially triggering instability in the region, as well as a robust commitment to building up Australia's long-range maritime strike and intelligence collection capabilities. Rudd's tell-tale reference in a private conversation with US Secretary of State Hillary Clinton that the United States may have to use armed force against China at some future point displayed a much harder edge to his government's thinking than was evident in public statements.[19]

The 2009 White Paper retained the implied sole purpose doctrine of earlier documents, which meant that Australian comments on US extended deterrence never received official US endorsement in a joint statement.[20] Nonetheless, it contained as strong a comment on the value of US extended nuclear deterrence as could be found in previous white papers:

> For as long as nuclear weapons exist, we are able to rely on the nuclear forces of the United States to deter nuclear attack on Australia. Australian defence policy under successive governments has acknowledged the value to Australia of the protection afforded by extended nuclear deterrence under the US alliance. That protection provides a stable and reliable sense of assurance and has over the years removed the need for Australia to consider more significant and expensive defence options.[21]

This public enthusiasm of the Rudd-Gillard-Rudd governments for continuing US extended deterrence was somewhat at odds with their concurrent strong endorsement of nuclear disarmament over the 2007–13 period. Lowy Analyst Rory Medcalf commented before a Parliamentary committee that:

> It is well and good for a country such as Australia to browbeat others about nuclear disarmament, but we do not live in as dangerous a neighbourhood as most of these other countries. However, we feel the need for an American nuclear umbrella. It is a challenge for our credibility on this issue.[22]

---

19 Anon., 'US Embassy Cables: Hillary Clinton Ponders US Relationship with its Chinese "Banker"', *The Guardian*, 5 December 2010.

20 Contrast this with a joint US-South Korean statement in 2009 that 'reaffirmed the US commitment to provide extended deterrence for the ROK, using the full range of military capabilities, to include the US nuclear umbrella, conventional strike, and missile defense capabilities'. For details, see 'The 41st US-ROK Security Consultative Meeting Joint Communique, October 22, 2009, Seoul', available at http://www.defense.gov/releases/release.aspx?releaseid=13072 [accessed 2 September 2014].

21 Australian Department of Defence, *Defending Australia in the Asia Pacific Century: Force 2030* (Canberra: Commonwealth of Australia, 2009), 50.

22 Comments by Rory Medcalf, Lowy Institute for International Policy, quoted in Commonwealth of Australia, 'Inquiry into Nuclear Non-proliferation and

The Gillard government captured this disjuncture, but hardly resolved it in the paragraph on deterrence in its own 2013 Defence White Paper:

> As long as nuclear weapons exist, we rely on the nuclear forces of the United States to deter nuclear attack on Australia. Australia is confident in the continuing viability of extended nuclear deterrence under the alliance, while strongly supporting ongoing efforts towards global nuclear disarmament.[23]

The submissions to the US NPR made by Australia's Department of Defence, however, underscored that Australia was even more deeply concerned, and more supportive of maintaining the effectiveness of nuclear deterrence, than was suggested by the government's public statements. It cautioned that 'The process of disarmament is likely to be complex and prolonged and must be pursued carefully while retaining a stable balance of nuclear postures that is sensitive to geopolitical realities,'[24] and contained several paragraphs – largely blackened out in the documents released under the FOI act – on 'assurance and deterrence in north Asia'.[25] The statement that 'Confidence that a nuclear attack on US allies would be met with a response-in-kind has assured very close US allies, like Australia, that they do not need to develop their own nuclear weapons'[26] clarified the somewhat cryptic comment in the public White Paper, and the submission called for the US government 'to ensure that [the Life-Cycle Extension Program (LEAP) and Stockpile Stewardship Program (SSP)] are properly funded and resourced'.[27] It acknowledged that Australia and the United States differed in regards to the use of nuclear deterrence against non-nuclear threats, but was remarkably cautious in commenting that 'If … the United States was disposed to make a "sole purpose" declaration, the Australian Government would be supportive in principle, but would want to be closely consulted on the specific details.' Rather than arguing for US adoption of Australia's position, the Australian submission recommended consideration be given to strengthening the 'negative security assurances' instead – a far less consequential change than that ultimately adopted in the NPR.[28]

Disarmament', *Parliamentary Joint Standing Committee on Treaties, Report no. 106* (Canberra: Commonwealth of Australia, September 2009), 17.

23 Australian Department of Defence, *Defence White Paper 2013* (Canberra: Commonwealth of Australia, Canberra, 2013), 29.

24 Australian Department of Defence, 'Draft Submission to the United States Nuclear Posture Review 2009', 2. Made available through the Freedom of Information Act at http://www.defence.gov.au/foi/Decisions/DisclosureLog.asp [accessed 12 July 2014].

25 Ibid., 3.

26 Ibid., 2.

27 Ibid., 5.

28 Ibid., 4.

Uranium Exports and Mining

Between 2007 and 2013, reference to a global 'nuclear renaissance' pervaded discussions over the most appropriate energy mix for countries in an era where climate change had emerged as a major global challenge. Evidence that many developing countries were moving to invest in the construction of new nuclear infrastructure reflected the strong expansion in China and India in particular of existing civil nuclear energy programs. If states were to achieve ambitious economic growth targets through base load power generation, while at the same time lowering their carbon emissions, investing in nuclear energy seemed to be an obvious pathway. This raised the spectre that countries new to the nuclear endeavour that were insecure about their regional environment could be tempted to invest in dual-use technologies with military applications down the track. Indeed, by definition, the very process of becoming proficient in managing the nuclear fuel cycle means that policy makers in some states will have enhanced strategic options in future.[29] The 2011 Fukushima disaster punched a hole in many of these nuclear ambitions, with several states in Europe winding back their nuclear programs as a result.[30] Ironically, with the obvious exception of Japan, the impact of Fukushima on nuclear investment was less obvious in Asia than in the rest of the world. Indeed, after undertaking high profile safety checks on their operating facilities, China and India made it clear they had no intention of pulling back from their commitment to expanding nuclear power.

Australia's engagement in the nuclear fuel cycle as a supplier of uranium thus continued to generate controversy and debate between 2007 and 2013. The 2007 election had seen opposing positions over whether Australia should export uranium to India. The Howard government's announcement that it would do so, despite New Delhi not being a member of the NPT, was opposed by the Labor Party, and Opposition Leader Kevin Rudd pledged that he would reverse John Howard's policy shift on uranium exports. This debate was significant because it signalled, at least for a limited time, the end of essential bipartisanship in Australian uranium export policy, which had made sales of yellowcake to non-nuclear weapons states contingent on recipients accepting full-scope nuclear safeguards. The ALP responded to the Howard government's move away from this cornerstone of Australian nuclear policy by defending the grand bargain and re-embracing the non-proliferation and nuclear disarmament

29 For analysis on this point, see Matthew Fuhrmann, *Atomic Assistance: How 'Atoms for Peace' Programs Cause Nuclear Insecurity* (Ithaca, NY: Cornell University Press, 2012), chapter 8.

30 See Helen Pidd and Suzanne Goldenberg, 'Germany suspends power station extension plans as nuclear jitters spread', *The Guardian*, 15 March 2011, available at http://www.theguardian.com/environment/2011/mar/14/germany-japan-nuclear-industry?guni=Article:in%20body%20link [accessed 2 August 2014].

rhetoric of the Hawke and Keating governments. Thus, not only did the Rudd-led opposition deem the India decision to be 'a very bad development' for non-proliferation, it also argued that the decision tarnished Australia's standing as a 'force for good on the NPT'.[31] Rudd argued in opposition that the Howard government's reliance on bilateral safeguards was insufficient to secure Australia's national interests beyond those of an economic nature, suggesting that export of uranium to India under such conditions would in fact undermine the country's long-standing interest in the health of the non-proliferation regime and potentially contribute to strategic instability in Asia:

> Mr Howard needs to get that right because fundamental national security interests for us and the next generation of Australians is at stake here. We respect the nuclear non-proliferation treaty and the International Atomic Energy Agency because we've got to prevent nuclear weapons proliferations in our region, our neighbourhood, our own backyard. No one in Australia wants a nuclear arms industry aided by US, in the Indian sub-continent or between India and China, because we have failed to ensure the upholding of the NPT.[32]

At the same time as it re-embraced the 'Labor tradition' on nuclear exports, the ALP at its April 2007 party conference overturned a long-standing element of this tradition by renouncing its 'three mines' policy. According to Rudd and other senior ALP figures at the time, this decision was taken in order to signal both the party's support for the expansion of uranium mining and its recognition of the potential economic opportunities presented by the expansion in demand for nuclear energy in the Asia-Pacific.[33] In the process of overturning this stance, the ALP also attempted to sever the connection between debates surrounding uranium mining, the safety of the nuclear fuel cycle and contemporary proliferation problems. Chris Evans, the then Shadow Minister for National Development, Resources and Energy, argued that the Labor Party had to 'acknowledge the reality of Australia's existing uranium industry and to recognise that we are but one of many suppliers of uranium'. Moreover, he conceded that 'an Australian ban on more uranium mining would not address

---

31 Australian Broadcasting Corporation (ABC Online), 'Rudd Slams Indian uranium decision', *ABC News*, 15 August 2007, available at http://abc.gov.au/news/stories/2007/08/15/2005697.htm [accessed 2 August 2014].

32 Australian Broadcasting Corporation (ABC Online), 'Labor Pledges to Overturn India Uranium Deal', *ABC News*, 17 August 2007, available at http://www.abc.net.au/news/stories/2007/08/17/2007382.htm [accessed 2 August 2014].

33 See for example Chris Evans, 'Labor and Uranium – An Evolution', speech by Senator Chris Evans, Leader of the Opposition in the Senate, Shadow Minister for National Development, Resources and Energy, to the Labor Business Roundtable, Parmelia Hilton, Perth, 23 March 2007, available at http://www.alp.org.au.; 'Kevin Rudd, TV Interview', *7:30 Report: ABC Television*, 30 April 2007, available at http://www.alp.org/media/0407/tviloo301.php?mode=print [accessed 2 August 2014].

the very serious issues of the nuclear industry and weapons proliferation.'[34] In that sense, the mining decision was consistent with the core assumption behind the bipartisan consensus in Australian politics since the late 1970s that only by becoming a supplier of uranium could Australia hope to exert any influence over the direction of the non-proliferation regime. More prosaically, it also represented part of Labor's counter to Prime Minister Howard's attempt to 'wedge' the Opposition politically in relation to the issues of uranium exports and climate change.

Once in power, Rudd acted on the ALP's pre-election statements with respect to the non-proliferation regime and uranium exports. As foreshadowed by the response to the Howard government's India deal, the new government's core approach was to return to the position of restricting sales of uranium to countries that were signatories to the NPT while actively pursuing 'more effective international export control regimes through the IAEA and tighter controls on the transfer of nuclear technology'.[35] The ALP had also pledged in August 2007 to reinvigorate Australia's 'active middle power diplomacy' in the non-proliferation sphere, the centrepiece of which was to be an Australian effort to 'establish and lead a new caucus of like-minded countries including both nuclear suppliers and nuclear users'.[36] In government, these commitments resulted in the repudiation of uranium sales to India in January 2008 and the establishment of the ICNND.[37]

Ultimately, the Rudd and Gillard governments' position on uranium exports and non-proliferation returned to the rhetorical style and content that had largely characterized Australian nuclear policy under all governments since Malcolm Fraser's. In August 2009, for example, then Foreign Minister Stephen Smith presented a justification for Australia's role as a uranium supplier that was remarkably similar to that presented by Fraser before Parliament in August 1977. Smith argued that because Australia possessed 'the world's largest uranium reserves', it had 'both the responsibility and standing to help ensure that international nuclear cooperation serves exclusively peaceful purposes and does not contribute to nuclear weapons programs'.[38] Furthermore, as a consequence

34 Evans, 'Labor and Uranium – An Evolution'.

35 Rudd, 'Leading, Not Following'.

36 Robert McClelland, 'A New Agenda for Australia in Nuclear Non-Proliferation and Disarmament', address to UN Association of Australia and Medical Association for the Prevention of War, Canberra, 14 August 2007, available at http://nautilus.org/apsnet/a-new-agenda-for-australia-in-nuclear-non-proliferation-and-disarmament/ [accessed 2 August 2014]. McClelland was then Labor's foreign affairs spokesman.

37 See 'Govt. reverses uranium sale to India', *ABC News*, 15 January 2008, available at http://www.abc.net.au/news/stories/2008/01/15/2139188.htm [accessed 2 August 2014].

38 The Hon. Stephen Smith, MP, 'Building Momentum: Australia, Nuclear Non-Proliferation and Disarmament', Sir Arthur Tange Lecture in Australian Diplomacy,

of this 'responsibility and standing', Australia had developed 'the world's strictest practices for safeguards for uranium supply', a policy, according to Smith, that 'becomes increasingly relevant as countries take up nuclear energy as a means to reduce greenhouse gas emissions and secure future energy supplies'.[39] Thus the Rudd government's position would be characterized by the assumption that the conditioning of uranium supply by the 'world's strictest safeguards' provided Australia with the best means of achieving its non-proliferation preferences.

Indeed, the Rudd government's handling of the brief controversy surrounding uranium exports to Russia in 2008 demonstrates that the idea of using uranium exports for policy leverage was very much alive and well in the Labor Party, despite the acknowledgment under Keating that the uranium export ban on France had failed to achieve any of Australia's objectives. As the Rudd government contemplated ratifying a uranium deal with Russia concluded by the Howard government in 2007, Russian military forces invaded Georgia from the disputed territory of South Ossetia. This prompted speculation that the Rudd government could signal its displeasure with events in the Caucasus by deciding not to ratify the agreement, and the Labor-dominated Joint Standing Committee on Treaties decided against recommending ratification of the agreement, largely on the basis of the Russia-Georgia conflict.[40] Again, the Australia Russia uranium deal was ultimately ratified by Rudd's successor, Julia Gillard, on the sidelines of the G20 meeting with Russian President Dmitry Medvedev in Seoul in November 2010[41] – with no evidence that either the delay, or signature, had any meaningful influence on Russian policy. In language uncannily similar to Gareth Evans' comments overturning the ban on exports to France in 1986, Gillard stated that exports to Russia would 'increase export opportunities and create jobs for Australian uranium suppliers and consolidate Australia's position as reliable provider'.[42]

---

12 August 2009, available at http://www.foreignminister.gov.au/speeches/2009/090812_tange.html [accessed 2 August 2014].

39 Ibid.

40 B. Packham, 'Nuclear deal with Russia put on ice', *The Herald Sun*, 19 September 2008, available at http://www.heraldsun.com.au/news/national/nuclear-deal-with-russia-put-on-ice/story-e6frf7l6-1111117524682 [accessed 2 August 2014].

41 J. Norman, 'New Risks from Australia-Russia Uranium Deal', *Online Opinion*, 19 November 2010, available at http://www.onlineopinion.com.au/view.asp?article=11251 [accessed 2 August 2014].

42 See 'Aust. Signs Off on Russia Nuclear Deal', *SBS News*, 11 November 2010, available at http://www.sbs.com.au/news/print/Article/1409941 [accessed 2 August 2014]; Stephen Johnson, 'Aussie Uranium Won't be Used in Russian Nukes, says Kevin Rudd', *Herald Sun*, 19 March 2010, available at http://www.heraldsun.com.au/news/breaking-news/aussie-uranium-wont-be-used-in-russian-nukes-says-kevin-rudd/story-e6frf7jx-1225842979265 [accessed 2 August 2014].

Despite ultimately making an exception for India in 2011, the final major new development during the Rudd-Gillard-Rudd period with respect to uranium export policy suggested that Labor retained faith in the basic assumptions underlying the 'grand bargain'. Indeed, the Gillard government was faced with domestic criticism in 2011 after it became public that it had begun discussions with the United Arab Emirates (UAE) regarding the conclusion of a bilateral nuclear safeguards agreement as a first step towards the conclusion of a uranium export deal. Australian Greens Senator Scott Ludlum argued that such a development would constitute 'very poor judgement'[43] and characterized government policy as a 'uranium free-for-all':

> Apart from North Korea, is there any country this government has ruled out as a uranium customer? Whatever assurances the UAE government gives, can we trust a regime that denies its people basic political and human rights?[44]

Foreign Minister Kevin Rudd, however, was quick to argue that the negotiation of a bilateral safeguards agreement was further proof of the 'strict non-proliferation conditions' that Australia required for the export of its uranium.[45]

Back and Forth on the India Deal

Although the controversy over uranium exports to India had extended across the terms of three Prime Ministers (i.e. Howard, Rudd and Gillard), the contours of the debate suggested that many within Australian politics continued to believe that Australian uranium remained, to use Fraser government Minister Ian Sinclair's phrase, a 'tangible incentive' capable of inducing 'good' non-proliferation behaviour among potential recipients. As detailed in the previous chapter, the Howard government had come to the conclusion that exporting uranium to India along the lines of the deal struck between Washington and New Delhi in March 2006 provided a non-proliferation dividend by ensuring that at least some of India's nuclear infrastructure came under the remit of

43 Enda Curran, 'Australia Eyes UAE Uranium Talks', *Wall Street Journal*, 9 March 2011, available at http://online.wsj.com/news/articles/SB10001424052748703560404576189570755638238 [accessed 2 August 2014].

44 Quoted in Dan Oakes, 'Emirates Gets Nod to Buy Uranium', *The Age*, 10 March 2011, available at http://www.theage.com.au/national/emirates-gets-nod-to-buy-uranium-20110309-1bo0m.html [accessed 2 August 2014].

45 Ibid. For an overview of the agreement see 'Australia-UAE Nuclear Cooperation Agreement', Media Release, Bob Carr MP, Minister for Foreign Affairs, 1 August 2012, available at http://foreignminister.gov.au/releases/2012/bc_mr_120801a.html [accessed 10 September 2014].

IAEA inspections.[46] However, the Rudd government argued, in a faithful reading of the grand bargain, that by withholding uranium sales Australia strengthened non-proliferation by not 'rewarding' India for remaining outside of the NPT and retained a specific bargaining chip to extract concessions in line with Australia's non-proliferation preferences.[47]

Nonetheless, the Rudd government confronted the difficult task of attempting to reconcile its stated desire to develop a 'strategic partnership' with New Delhi with its agenda to reinforce Australia's traditional approach to uranium exports and non-proliferation. This dilemma was illustrated most starkly in September 2008 when the Nuclear Suppliers Group (NSG) agreed to exempt India from the Group's ban on nuclear exports to countries that do not have IAEA 'full scope' safeguards in place. Despite its decision to overturn the 2007 uranium deal, the Rudd government acceded to the NSG decision. According to then Foreign Minister Stephen Smith, this decision was taken with 'consideration of the strategic importance of the issue for both India and the United States' and India's formal declaration that it would stand by its non-proliferation commitments and uphold its moratorium on nuclear tests.[48] Smith was then careful to note that the NSG's decision 'will enable civil nuclear supply to India by those NSG participating nations who choose to do so', before re-stating that it was the Rudd government's choice not to supply India with uranium on the basis of 'long standing' policy.[49] This position left the Rudd government open to charges that it was pursuing a contradictory, even hypocritical, policy or one defined by confusion and indecision. Indeed, Opposition Foreign Affairs spokesman Andrew Robb asserted that the decision to reverse the Howard-era agreement was 'putting grubby party politics ahead of the national interest',

---

46 For this view see for example the Hon. Alexander Downer, MP, 'Speech and Question and Answer Session, Energy Environment and Air Quality Policy Forum', Los Angeles, 13 January 2007, available at http://www.foreignminister.gov.au/transcripts/2007/070113_eeaqpf.html [accessed 9 September 2014]; Andrew Robb, 'Uranium Sales to India: A Strategic Imperative', *The Sydney Papers* (Autumn 2008): 55–62.

47 See for example the Hon. Stephen Smith, MP, 'Australia and India: A New Partnership in the Asia Pacific Century', address by the Minister for Foreign Affairs and Trade to the Indian Council of World Affairs, New Delhi, 11 September 2008, available at http://www.foreignminister.gov.au/speeches/2008/0809011_newdelhi.html [accessed 10 September 2014].

48 Stephen Smith, 'Australia and India: A New Partnership in the Asia Pacific Century', address by the Minister for Foreign Affairs and Trade to the Indian Council of World Affairs, New Delhi, 11 September 2008, available at http://www.foreignminister.gov.au/speeches/2008/0809011_newdelhi.html [accessed 13 November 2010].

49 Ibid.

suggesting that the decision had more to do with internal ALP factional politics than promoting Australia's national interests.[50]

Australia's position was placed under further scrutiny during the Gillard government, following the Obama administration's commitment in November 2010 to support India's bid to become a full member of the NSG.[51] International developments appeared to be making Australian policy redundant. There was next to no chance that India would unilaterally forgo its nuclear weapons capability in order to sign the NPT as a non-nuclear weapons state, in an environment where its nuclear status was being accepted by an increasing number of other countries and nuclear supplies. In reality, standing apart from the India deal threatened to create an enduring obstacle to closer Australian relations with New Delhi, for no apparent gain in non-proliferation outcomes. This was all the more galling to India due to the fact that in voting for the NSG waiver in September 2008, Foreign Minister Stephen Smith cited the same factors as those identified by the previous Howard government to justify its decision to export uranium to India: i.e. India's 'positive' record on non-proliferation; the non-proliferation gains of placing some of India's civil nuclear industry/infrastructure under IAEA inspections; and India's status as a rising power.[52]

A combination of inertia and domestic political considerations stifled reconsideration of Australian policy on this issue between 2008 and 2011.[53] The outlines of this debate, however, were altered in November 2011. In an opinion piece in *The Age,* Prime Minister Gillard not only called on her party to revisit the issue of uranium sales to India at the forthcoming ALP National Conference in December 2011 but also expressed her personal opinion that 'It is time for Labor to modernise our platform and enable us to strengthen our connection

---

50 Andrew Robb, 'Nuclear India is Good for the Globe', *The Australian*, 28 July 2008, available at http://www.theaustralian.com.au/news/opinion/nuclear-india-is-good-for-globe/story-e6frg6zo-1111117030788 [accessed 13 November 2013].

51 Geoff Hiscock, 'Scrutiny for Australia's Uranium Ban', *The Australian*, 11 November 2010, available at http://www.theaustralian.com.au/business/news/scrutiny-for-australias-india-uranium-ban/story-e6frg90o-1225952118083 [accessed 13 November 2010].

52 Stephen Smith, 'Australia and India: A New Partnership in the Asia Pacific Century', address by the Minister for Foreign Affairs and Trade to the Indian Council of World Affairs, New Delhi, 11 September 2008, available at http://www.foreignminister.gov.au/speeches/2008/0809011_newdelhi.html [accessed 13 November 2010].

53 See for example Rory Medcalf, 'Powering Major Powers: Understanding Australian Uranium Export Decisions on China, Russia, and India', in Michael Clarke, Stephan Frühling and Andrew O'Neil, eds, *Australia's Uranium Trade: The Domestic and Foreign Policy Challenges of a Contentious Export* (Farnham: Ashgate, 2011), 180–81.

with dynamic, democratic India.'[54] Such a 'modernisation' of policy would be made on the basis of a potpourri of now familiar arguments: India was a democracy; uranium exports to India would 'increase jobs'; and Australia would hold India to 'the same standards we do of all countries for uranium export – strict adherence to IAEA arrangements and strong bilateral undertakings' to ensure that its uranium would only be used for 'peaceful purposes'.[55] While senior ministers such as Treasurer and Deputy Prime Minister, Wayne Swan, and Foreign Minister, Kevin Rudd, were reportedly supportive of the Prime Minister's position, others, primarily those on the Left of the party remained opposed on the basis of long-standing arguments. Backbench MP Gavin Marshall perhaps best summed up these arguments when he asserted that, 'It's irresponsible to sell uranium to a nuclear-armed country that has not signed the nuclear non-proliferation treaty and is also in regular conflict with another nuclear-armed power.'[56]

These two basic lines of argument were in evidence during debates at the ALP's subsequent December 2011 National Conference. Despite 'feisty' debate, however, the Prime Minister prevailed, 206 votes to 185 on the floor of the conference.[57] Although there was much commentary that either lauded or lamented this decision, it had become abundantly clear that the international context had crucially influenced policy direction.[58] The response of the Minister for Defence, Stephen Smith, to questioning after the conference vote was revealing in this respect. Asked whether he had any qualms or fears of the potential consequences of uranium exports to India, he replied:

> Absolutely not. I think this is a deeply significant decision. This will advance Australia's interests. What changed the nature of this discussion was when India

54 Julia Gillard, 'An Opportunity for Sound and Fury to Signify Something', *The Age*, 15 November 2011, available at http://www.theage.com.au/opinion/an-opportunity-for-sound-and-fury-signifying-something-20111114-1nfi3.html [accessed 9 June 2014].

55 Ibid.

56 Phillip Coorey and Ben Doherty, 'Labor splits over decision on uranium', *The Age*, 16 November 2011, available at http://www.theage.com.au/environment/energy-smart/labor-splits-over-decision-on-uranium-20111115-1nhab.html [accessed 9 June 2014].

57 Katherine Murphy, 'Labor Votes in Favour of Selling Uranium to India', *The Age*, 4 December 2011.

58 For divergent reactions to the ALP's debates see Rory Medcalf, 'Uranium Sales to India Would Spread Trust, Not Nuclear Arms', *The Age*, 2 December 2011, available at http://www.theage.com.au/action/printArticle?id=2811203 [accessed 9 June 2014]; Marianne Hanson, 'Selling Uranium to India a Blow to Arms Control', *Canberra Times*, 6 December 2011, available at http://www.canberratimes.com.au/news/opinion/editorial/general/selling-uranium-to-india-a-blow-to-arms-control/2382464.aspx [accessed 9 June 2014].

> entered into the India-United States Civil Nuclear Agreement in 2007. That agreement was approved by the International Atomic Energy Agency and the Nuclear Suppliers Group over the course of 2008–2009. That effectively put India under the international nuclear regulators for the first occasion and India gave a series of undertakings including a moratorium on future nuclear testing, signing up the International Atomic Energy Agency's additional protocol, splitting its civilian nuclear technology from the military program. That effectively gives you the same protections that you'd get if a country signed the Nuclear Non-Proliferation Treaty, which of course has been a stumbling block for many years so far as India is concerned.[59]

Thus, the Gillard government had come to the same policy position on the basis of the same logic that the Howard government did when it approved in-principle uranium exports to India in August 2007. The logic of this position was that, in a situation where it is inconceivable that India would sign the NPT as a non-weapons state, some international transparency via IAEA inspections and the implementation of bilateral safeguards is better than none at all. At the same time, however, it also implicitly acknowledged the geostrategic considerations that had led to the deal in the first place. For Smith:

> this has been the best way of making sure that India, as it emerges as a superpower, as it takes its rightful place as the world's largest democracy and one of the three countries that will be superpowers in this century – the United States, China and India – has taken its rightful place and has voluntarily agreed to go under the governance of the International Atomic Energy Agency. That's a good thing. That's progress. That's improvement.[60]

In the end, the Gillard and Howard governments' agreement that Australia should recognize India as a de facto nuclear weapons state thus also harked back to the debates of pre-NPT days. Then, as now, the 'Menzian' view was that ultimately, Australia's interest was in making nuclear weapons a preserve of the great powers – and in 2011, both parties in Australia agreed that India was one of these.

Conclusion

By the time the Rudd government was defeated at the September 2013 national election, the settings of Australian nuclear policy had returned to a high degree

---

59 Stephen Smith, MP, Minister for Defence – Interview with Virginia Trioli, *ABC News Breakfast*, 5 December 2011, available at http://www.minister.defence.gov.au/2011/12/05/minister-for-defence-interview-with-virginia-trioli-abc-news-breakfast [accessed 9 June 2014].

60 Ibid.

of bipartisanship. While the first Rudd government sought to reverse some of the policy settings laid down by its predecessor, between 2010 and 2013 Julia Gillard's government took an approach that had more in common with the Howard government than with the government led by Kevin Rudd between 2007 and 2010. Caution over carefully balancing rhetorical commitment to nuclear disarmament with a perceived need for reassurance from US extended deterrence and a more permissive attitude regarding uranium exports to India were key hallmarks. Kevin Rudd's commitment to infusing Australian nuclear policy with greater Labor 'tradition' was captured in the ICNND initiative and the decision to oppose uranium sales to India, but there was in reality little wriggle room for his government to radically reshape directions – nor did his government have any real intention to challenge US nuclear policy, as was evident from the remarkably cautious submission to the 2010 US NPR.

The 2007–13 period thus confirmed that Australian nuclear policy has never really been driven by partisan forces. Policy directions and priorities over time have instead been shaped to a remarkable degree by agreement between the major parties, and bipartisan agreement was never greater than at the end of the Rudd-Gillard years. At the same time, however, it is also hard to argue that this consensus did not include a significant element of stagnation. Despite the high-profile ICNND, in the end neither the Rudd nor the Gillard government contributed much of substance that was new to Australia's nuclear policy overall.

Conclusion
Australian Nuclear Policy in the 20th and 21st Century

Australia was very much 'present at the creation' of the nuclear age in 1945. As a supplier of uranium to the Anglo-American nuclear weapons program, Australia gained an early foothold in deliberations about the nuclear fuel cycle in international relations. In the aftermath of the Second World War, the country's own approach to nuclear matters reflected a sense of optimism and ambition that the nuclear fuel cycle could be readily exploited for national development purposes. The economic potential of the atom from the perspective of policy makers dovetailed closely with the view that Australia should preserve the technological resources to reach a threshold nuclear weapons capability if the security environment in Southeast Asia deteriorated markedly. Contrary to common misconception, Australian governments did not endeavour to acquire an indigenous nuclear weapons capability at any point; but they did seek to keep the option open right up until the shelving of the Jervis Bay project in 1971. Indeed, much of Australia's nuclear policy between 1945 and 1972 was characterized by an attempt to keep the country out of international commitments that were perceived as having the potential to constrain Australia's nuclear options down the track.

Ever since Australia acceded to the NPT as a non-nuclear weapons state in 1973, Australian nuclear policy has been a delicate balancing act between the strategic, economic and normative domains. Australia supported the new NPT regime and saw benefits for its own security. The possibility that it might have to rethink the nuclear weapons option, however, should strategic circumstances deteriorate, was not yet seen as unthinkable. Successive governments have attempted to reconcile the objective of slowing the nuclear arms race between the superpowers, with supporting the US nuclear deterrent as a source of global stability. Australia seeks to limit and reverse nuclear proliferation to other countries, but continues to rely on extended nuclear deterrence from the United States. It seeks commercial benefit from the export of uranium and, on occasion, uses it to build strategic relationships in Asia, while at the same time trying to limit the spread of nuclear technology and nuclear latency of regional countries.

Throughout the second half of the Cold War, Australian nuclear policy attempted to achieve these core goals in a relatively stable international strategic context, defined by a logic of nuclear deterrence that was dominated by the competition between the two superpowers and limited global access to nuclear technology. In the immediate post-Cold War era, Australia was able to help bring about a number of important steps in non-proliferation – including the adoption by the Nuclear Suppliers Group of a full-scope safeguards condition for nuclear exports, the indefinite extension of the NPT, the development of new safeguards technologies and the IAEA Additional Protocol, and the agreement on a Comprehensive Test Ban Treaty. After the 9/11 attacks in 2001, the advent of mass-casualty terrorist groups seemed to undermine confidence in the rationality-based logic of deterrence. This reinforced fears about the nuclear weapons programs of 'rogue states' including North Korea, Iran and Iraq, while transnational nuclear technology and components networks have highlighted the limitations of export-control regimes regulating nuclear supply. Like other countries around the world, Australia therefore paid closer attention to the practical aspects of nuclear safety and security, and the interception of illicit trafficking of nuclear technology.

Australia's policy with respect to nuclear issues has thus evolved significantly since the dawn of the atomic age in 1945. Few policy makers of the 1950s and 1960s would recognize the contours of Australia's approach to nuclear issues today. In particular, the depth of Australian engagement in global forums dedicated to stopping the spread of nuclear weapons and promoting non-proliferation as a popular norm in international relations bears little relationship to how Australian decision makers thought about nuclear policy before the 1970s. Indeed, while Australia began to deepen its involvement in multilateral non-proliferation and disarmament architecture after it ratified the NPT in 1973, it was not until the late 1980s that Australia became recognized as a middle power norm entrepreneur in the field of arms control and disarmament. The loosening of Cold War shackles, coupled with increasing ambition to play a genuine global role in general foreign policy terms, meant that by the 1990s Australia was pursuing one of the most active diplomatic agendas of any country in the international system with respect to nuclear arms control and disarmament. While this has ebbed and flowed since, Australian governments from both sides of the political spectrum remain committed to this active diplomatic role.

Characteristics of Australia's Nuclear Policy

Six principal findings emerge from the analysis contained in this book. The first is that there has been a high degree of continuity in Australian nuclear policy since the Fraser government laid down the parameters for the export of uranium in 1977. Strict adherence to the NPT; a commitment to the application of full-scope safeguards over Australian nuclear transfers; and diplomatic efforts to strengthen the international non-proliferation regime at all levels have formed the framework for Australia's nuclear policy under successive Coalition and Labor governments. Despite some inevitable variations in emphasis, since 1977 governments of all political shades have endorsed this framework as the reference guide for developing nuclear policy.

The second key finding is that, from a historical perspective, a high level of bipartisanship has prevailed in shaping how successive governments have formulated the actual details of Australian nuclear policy. Unlike Canada, for example, it never developed nuclear reactors to capitalize on its uranium ore deposits. Even its uranium exports never fully lived up to their potential, as Australia's market share continues to lag well behind its share of worldwide deposits. Despite the strong persistence of vocal opposition to Australian uranium exports, no Australian government has seriously considered halting uranium sales. Moreover, in the absence of both nuclear power reactors and nuclear weapons stationed on Australia's own soil, much of the anti-nuclear policy pressure over the years has focused on the issue of uranium exports. The policy, bureaucratic and legislative attention to this topic by successive governments over several decades is quite remarkable, and ultimately anti-nuclear movements have thus had an effect of reinforcing constructive bipartisanship in Australia's nuclear policy. On the one occasion when a split has emerged between the Labor and Coalition sides over uranium sales following the Howard government's decision to back exports to non-NPT member India, the Gillard government eventually affirmed this policy shift in 2011. To say there were no differences between the major parties over nuclear policy is an exaggeration, but by 2013 the wheel had turned full circle for Australian nuclear policy. The prospect of witnessing further 'great debates' about Australia's engagement in the nuclear fuel cycle seemed lower than at any stage since 1945.

The third principal finding is that rather than damaging or inhibiting Australia's credibility with respect to nuclear diplomacy, extended nuclear deterrence as part of the bilateral security alliance with the United States has provided reassurance for successive governments that has contributed to a degree of confidence in Australia's international nuclear engagement that may not have otherwise been present. While governments have not pursued arms control and disarmament initiatives antithetical to America's global nuclear commitments – itself hardly a revelation given the nature of the formal alliance between Australia and the United States – there is scant evidence that Australian governments have seen the US alliance as being at odds with promoting arms control and disarmament, even when Washington has expressed ambivalence about Australia's approach. It is worth asking, however, whether successive Australian governments would have been as confident in the NPT and the non-proliferation regime without the reassurance of extended nuclear deterrence. Although a counterfactual question, the authors believe the answer to be in the negative.

The fourth key finding is that unlike a number of other similarly positioned middle powers, Australia has never seriously questioned the legitimacy of the nuclear arsenals of the five NWS recognized under the NPT. This tendency reflects the influence of the 'Menzian' vision of nuclear weapons that perceives nuclear weapons to be a stabilizing factor in international politics – particularly through the mechanism of deterrence – as long as they are wielded by 'responsible' great powers. This vision has at times been challenged, most notably during the Keating years, by a 'disarmer' vision that sees nuclear weapons as 'order destroyers' rather than 'order builders' and thus inimical to Australian strategic and security interests.[1] Despite this challenge, successive Australian governments of both major political persuasions have ultimately structured much of the country's nuclear policy around this view. The Howard, Gillard and now Abbott government's acceptance of India as an NWS also speaks to the endurance of this vision in Australian thinking about the role of nuclear weapons in international politics.

The fifth key finding is that Australia's commitment to the global non-proliferation regime has been crucially shaped by an enduring national security interest in limiting nuclear proliferation in its immediate region. This interest was a major factor in Australia's deliberations regarding membership of the NPT in the late 1960s and early 1970s, with policy makers focusing on the role of nuclear proliferation in South East Asia in presenting arguments both for and against Australian accession to the Treaty. Australia's decision to sign and ratify the Treaty as an NNWS, and its subsequent support for the development of strict nuclear safeguards and consolidation of the non-proliferation regime, were seen as important steps to proliferation-proof Australia's region. As noted in Chapter 4, the genuine concern with the strategic consequences of regional nuclear proliferation stood in contrast to other Western US allies who perceived the NPT as a mechanism to manage the Cold War, and was crucial for Australia's diplomacy on issues relating to the NPT as well as the CTB under the Fraser government and its successors. Australia's global non-proliferation diplomacy was always complemented with a less prominent, but equally important and telling, focus on the practical barriers to proliferation in South East Asia, including for example technical support to regional nuclear institutions and authorities since the 1970s, the Hawke government's support for SPNFZ in the 1980s, the Keating government's reticence on the US 'counterproliferation' agenda in the 1990s, and the 'Asia-Pacific Safeguards Network' and Australian advocacy of PSI in the 2000s. Australian non-proliferation interests lie in stable global regimes, but in the end these regimes are a means to achieve security for Australia in its own region.

The sixth key finding of this book is that successive Australian governments since the 1960s have been ambivalent towards suggestions that Australia become

1 Rod Lyon and Christine Leah, 'Three Visions of the Bomb: Australian Thinking about Nuclear Weapons and Strategy', *Australian Journal of International Affairs* 64, no. 4 (2010): 451–2.

more technologically engaged in the nuclear fuel cycle, including considering becoming a consumer of nuclear power. Although it has eschewed nuclear power production, Australia built a sizeable nuclear research establishment in the 1950s and 1960s. The Australian Atomic Energy Commission operated uranium centrifuges in the 1970s and into the 1980s, and ANSTO to this day researches technologies that are relevant to reactor and fuel cycle design. However, the closest Australia has come to having a serious domestic debate about over these issues in recent years occurred under the Howard government. The 2006 UMPNER report ('Switkowski report') canvassed in detail uranium enrichment and civilian nuclear power options for Australia, but failed to garner political support and was effectively shelved when the Rudd government entered office the following year. While Labor has a well-established track record since the 1970s of opposing nuclear power and the domestic enrichment of uranium, the Coalition too is keen to define discussion of Australia's nuclear fuel cycle engagement to uranium exports. When questioned about nuclear energy, the Abbott government's Industry Minister recently responded that:

> We are agnostic in terms of energy sources and I emphasise that the Abbott Government has no intention of leading a debate on nuclear power in Australia ... there is no need to have a debate in regard to nuclear energy in Australia but we should focus on the opportunities that nuclear energy presents in other countries and build our uranium industry to take advantage of that.[2]

Australian Nuclear Policy and the US Alliance

In light of these findings, it is clear that the orthodox narrative of Australian foreign and nuclear policy, of Australian emancipation from its reliance on 'great and powerful friends' (in relation to foreign policy in general) and the internalization of arms control and non-proliferation norms (in relation to nuclear policy in particular) after the early 1970s underplays significant continuities in Australia's approach to nuclear weapons and nuclear proliferation before and after the accession to the NPT. The conditions that allowed and fostered the increasing diplomatic activism of Australian governments in international non-proliferation are not only domestic, since the international context for nuclear policy also changed radically between the 1950s and 1960s on the one hand, and the 1970s and later decades on the other hand. The development of the NPT regime was in large measure due to US policy and pressure on its allies. In the end, a stable nuclear order was a goal that was shared by most industrialized

2 Australian Broadcasting Corporation (Online), 'Macfarlane Rules Out Nuclear as Report Considers Energy Future', *The World Today*, 23 September 2014, available at http://www.abc.net.au/worldtoday/content/2014/s4092959.html [accessed 9 June 2014].

countries, even if there were at times disagreements on how to strike the balance between economic and non-proliferation interests in the safeguards system. And the post-Cold War advances in arms control and non-proliferation occurred in the geostrategic honeymoon of the 1990s, during which time major multilateral agreements were concluded in a range of international policy areas. In that sense, Australian nuclear policy was, in its overall direction, following in the wake of US policy, rather than striking its own course.

The US alliance that bound Australia's security to considerations about global stability among the great powers thereby permitted Australia to shelve consideration of developing an indigenous nuclear weapons capability. This, in turn, enabled it to adopt a position of moral leadership on non-proliferation on the basis that it was a state that had once considered but ultimately rejected the nuclear weapons option. Thereafter it was sufficiently removed geographically from the nuclear weapons issues of the Cold War to present its non-proliferation diplomacy to other members of the international community as that of a 'good international citizen' seeking to strengthen the efficacy of the non-proliferation regime.[3] Insofar as geostrategic location is a virtue, Australia was (and remains) a virtuous country indeed.

And yet, whereas flirtations with a sovereign nuclear capability, especially under Prime Minster John Gorton in the 1960s, were based on a very narrow view on the benefits and feasibility of extended deterrence through the US alliance, the same can also be said of later vocal Australian advocacy for disarmament since the 1990s. It too rests on a relatively narrow view of nuclear weapons and nuclear deterrence, insofar as Australian politicians, officials and civil society activists generally see the purpose and role of extended nuclear deterrence as solely to deter other nuclear threats.[4] In that sense, the alliance has given Australia a relative sense of security, without exposing it to many of the dilemmas of nuclear strategy that confronted US allies in Europe and Northeast Asia. And yet, it is also clear that all Australian governments from Fraser to Rudd-Gillard have been very conscious of Australia's unusual situation in this regard, and have paid close attention to the stabilizing effect of US extended nuclear deterrence in the Northern hemisphere that, ultimately, also enhanced Australia's own security. In that sense, it is clear that the 'Menzian' consensus did not include the notion that non-nuclear weapons states should not benefit from nuclear deterrence, and the fact that Australia had only limited need for nuclear extended deterrence itself did not mean that Australian governments (of either side of the political spectrum) were keen to push for a de-valuation of extended nuclear deterrence after the Canberra and ICNND commission reported their recommendations.

---

3 As argued by Hanson and Ungerer, 'The Canberra Commission: Paths Followed, Paths Ahead'.

4 Lyon, 'Australia: Back to the Future?'

The Importance of the 'Grand Bargain'

What emerges then as the main, perhaps singular, change in the history of Australia's nuclear policy is not the accession to the NPT itself – which was after all conditional and openly argued to be such in government until the early 1980s. Rather, it was the 'grand bargain' that arose from the Fraser government's development of its uranium policy that was a radical break with earlier policy, and set in place the ideas that guided all governments that would follow in subsequent decades. It was based on the judgment that Australia's security would ultimately be enhanced, not limited, by a functioning global non-proliferation regime that helps keep nuclear weapons from its own region – a judgment that already underpinned earlier arguments for Australia's accession to the NPT. It was, however, when the Fraser government went beyond this and sought to employ Australia's activities across the whole nuclear field – most prominently, uranium exports, but also technical assistance programs, safeguards support and arms control diplomacy, especially in the CD – in support of this objective, that Australia's nuclear policy entered a new (and bipartisan) era.

The 'grand bargain' was only made operable by the specific dynamics of the international nuclear environment of the mid-to-late 1970s. In this regard the strategic context by the late 1970s was defined by a global logic of strategic restraint on behalf of the five NPT-recognized NWS, particularly the US and the Soviet Union. Meanwhile Australia's alliance with the US also not only provided Australia with an 'insurance policy' in the event of global or regional conflict, but also provided it with the necessary diplomatic 'space' to construct an activist approach to non-proliferation.

The interaction of such strategic factors with the economic situation of the era also provided Australia with incentives to ultimately strike its 'grand bargain'. The discovery of three of the world's largest untapped reserves of uranium in the Northern Territory and South Australia in the early 1970s combined with the oil shocks of the same period to convince Canberra that it could effectively leverage its uranium endowment for diplomatic gains. The 'oil crisis' stimulated renewed interest and demand for nuclear energy as a possible route to energy security with a subsequent projected rise in demand for uranium fuel to power nuclear reactors. Concern about availability of uranium fuel prompted consideration of the rapid development and use of plutonium fuel – through the operation of fast, or 'breeder', reactors – that would permit countries to become self-sufficient in nuclear fuel. This projected turn towards fast breeder reactor technologies, however, also held the potential to place nuclear weapons within reach of those states that were embarking on fast breeder reactor programs.

The proliferation threat of this so-called 'plutonium economy' was brought into stark relief with the Indian 'peaceful nuclear explosion' (PNE) of 1974 – whereby India used plutonium derived from a Canadian-supplied research reactor to acquire the fissile material for the development of a nuclear weapon. US non-proliferation policy under the Ford and Carter administrations thus began to place greater focus on the control of nuclear technology. This provided Australia with an opportunity to portray its decision to approve the export of uranium as a major contribution to international non-proliferation efforts. By only exporting uranium to non-nuclear weapons states if they were NPT member states and accepted 'stringent' bilateral and IAEA safeguards, Australian uranium export policy, it was argued, helped strengthen the non-proliferation regime by providing a tangible benefit – i.e. access to Australian uranium – to states that abided by the treaty.

While Australia's alliance with the United States provided Australia with an 'insurance policy' in the event of global or regional conflict it also provided it with diplomatic 'space' to construct an activist approach to non-proliferation – a dynamic particularly evident in Australia's position on CTB negotiations during the Fraser years. Finally, the grand bargain was very clearly struck in the context of the Carter administration's initiatives to halt the spread of the 'plutonium economy' by placing Australia in the best possible position to become a leading supplier of safeguarded uranium should the Carter administration's non-proliferation policy succeed. However, the outcome of the INFCE, in particular, demonstrated the failure of the administration's attempt to control the international nuclear fuel cycle by constraining access to reprocessing technology. This presented Australia with a major dilemma: with the INFCE deciding that nuclear fuel cycle technologies could not be ranked in terms of proliferation risk, what incentive remained for potential consumers of Australia's uranium to accept its 'stringent' safeguards and preconditions of supply? This dilemma would prove to be an enduring question for future Australian governments.

In that sense, the 'grand bargain' was the basis on which Australia would participate in the continuing development of the non-proliferation regime, and not a solution to the basic trade-offs inherent in nuclear policy. The four 'cornerstones' of the Fraser government's uranium export policy – strict adherence to the NPT, IAEA safeguards, bilateral agreements between Australia and importing countries and Australian participation in multilateral efforts to strengthen safeguards and the non-proliferation regime – established a position that Australia would couple a commitment to the NPT as the centrepiece of the international non-proliferation regime with an increasingly activist role in multilateral non-proliferation regimes. The conditions under which Australia's 'grand bargain' was struck demonstrated the linkages across the three domains of nuclear affairs – strategic, regime and economic. Yet they also suggest the contingent nature of the 'grand bargain' – i.e. it was tailored to a set of specific international and domestic conditions that prevailed at a particular historical moment. What emerges as the central question for the future of Australia's nuclear policy in the 21st century is thus whether the domestic and international conditions that gave rise to the 'grand bargain', and which have remained remarkably stable over the intervening decades, will remain in place as the global international system sheds many vestiges of the old, Cold War order.

The Future of Australia's Nuclear Policy

Much scholarship on Australia's nuclear policy in recent years has focused on the factors that led Australia to eschew the nuclear weapon option, with a view to making predictions about the future nuclear policy of Australia (and other countries) as the international order adapts to changes in the balance of power in the 21st century.[5] This book's findings about the enduring centrality of the ideas of the 'grand bargain', however, suggest some caution with historical extrapolations that ignore the development of (bipartisan) policy in the intervening decades. The suggestion that radical changes, if radical enough, may lead Australia to fundamentally revisit the nuclear weapons option,[6] is probably true, but also borders on the tautological. What matters for assessments of Australia's future policy is more the propensity of Australia to change course, and here the 'grand bargain' is important because it provides a narrative and approach that transforms Australian governments' views of themselves from being passive bystanders, to active participants with real influence on the shape and development of nuclear order. The extent of this influence in practice may be debatable – and this book has demonstrated that international context was generally far more important, and certainly essential, for Australia's successes in non-proliferation diplomacy – but the fact that the idea of Australian influence in the nuclear sphere has taken deep bipartisan roots in Australia is evident from the history of recent decades. It is this idea, rather than norms directly, that Australian governments seem to have internalized to a remarkable extent.

What lessons then flow from the analysis in this book for the future of Australia's nuclear policy? The first and most important one is that the 'grand bargain' is not a static policy construct, but one that Australian governments have had to adapt as new developments in the international system, non-proliferation regime or nuclear technology have thrown up both challenges and opportunities for Australian influence. What these will be in the future only time will tell, but it is likely that the issue of uranium enrichment in the context of international guarantees for the supply of nuclear fuel will, once again, be

5 See for example Luke Auton, 'Opaque Proliferation: The Historiography of Australia's Cold War Nuclear Weapons Option', *History Compass* 11, no. 8 (2013): 561–72.

6 See Stephan Frühling, 'Never Say Never: Considerations about the Possibility of Australia Acquiring Nuclear Weapons', *Asian Security* 6, no. 2 (2010): 146–69.

among them.[7] At the same time, it is important that Australian governments also accept the 'grand bargain' as a work in progress, indeed a challenge to find ways to further Australian influence in the non-proliferation regime. As this book has outlined, the decision to export uranium to India was at least not inconsistent with the spirit of Australia's nuclear policy since the Fraser years – but an agreement that allowed the export of uranium without the strict conditions that Australia has placed on all of its exports since that time,[8] certainly would be.

That said, there is unlikely to be any radical shift in Australia's stance on uranium exports, and given the consistent relative position of uranium in the league table of mineral exports (roughly 10% of coal exports and 3% of iron ore exports), it will never be a major economic consideration for governments. Variable public support for uranium exports means that all governments will continue to tread carefully in this area. Australian governments will in all likelihood remain committed to the view – exemplified by Minister Macfarlane's comments above – that the extent of Australia's involvement in nuclear energy should be confined to uranium exports. Barring a truly radical change in the environmental debates about the advantages and disadvantages of nuclear energy in Australia, governments will do their best to define this as an 'offshore' issue, not one that impacts on domestic politics and not an issue that will influence Australia's own future energy mix.

Just as the idea that nuclear power will play a role in Australia's future national development appears to be a thing of the past, so too is the notion that a future government will revise Australia's status as a non-nuclear weapon state. This is not because Australian governments ever formally chose to forego that option once and for all, rather that they might have 'internalized' this 'norm'. As shown in this book, well into the Hawke government, Australian ministers and their advisors still discussed the possibility that Australia might have to one day leave the NPT for security reasons. Rather, despite these discussions, Australia's geographic position is such that it never actually faced the kinds of existential threats that countries turn to nuclear weapons for, at least not since the dark days of 1942. Australian and European policy makers questioned the credibility of US nuclear guarantees in the 1960s, but it was only the latter which in fact relied on these guarantees in actual rather than hypothetical situations. Although the Fraser cabinet was concerned about the central balance, its concerns were primarily due to possible second-order effects

7 See for example John Carlson, 'The Potential for a Regional Uranium Enrichment Centre in Australia', *Security Challenges* 9, no. 4 (2013): 1–9.

8 John Carlson, 'Is the Abbott Government Abandoning Australia's Nuclear Safeguards Standards for India?', *The Interpreter*, 1 October 2014, available at http://www.lowyinterpreter.org/post/2014/10/01/Is-the-Abbott-Government-abandoning-Australias-nuclear-safeguards-standards-for-India.aspx [accessed 2 October 2014].

for Australia's security in its own region. To make it conceivable that future Australian governments would acquire nuclear weapons, this would require a demonstrable and direct security benefit that could only arise if major threats to Australia arose from its immediate neighbourhood. Hence, it would take a truly tectonic change in Australia's security circumstances for the acquisition of nuclear weapons to arise as a serious conversation among elites.[9]

As in previous decades, Australian governments should therefore pay particular attention to the way the global non-proliferation regime is rooted in Australia's own, Southeast Asian neighbourhood. Even so, the reassurance that is derived from having an alliance with the world's most powerful nuclear weapons state, and the argument that supporting, rather than further weakening a global regime would be the best response to regional proliferation, will almost certainly continue to reinforce Australia's non-proliferation restraint irrespective of how the security environment in Asia evolves in coming decades. Australia's high profile diplomatic role on issues pertaining to nuclear arms control and disarmament is central to the national role conception held by policy makers of Australia as a constructive middle power committed to good international citizenship,[10] but it is also based on a realistic appreciation of the country's geostrategic strategic situation.

While Australia as a champion of non-proliferation diplomacy looms large in any analysis of Australian nuclear policy, it is important not to lose sight of the key role Australia's technical capacity building role in its own region has played in promoting its non-proliferation interests. Australia has a well-established, if less conspicuous, tradition of technical contributions to practical problems that arise in non-proliferation. This includes Australia's Safeguards Support Program cooperation with the IAEA, participation in the Hexapartite Safeguards Project and the analysis of IAEA samples by ANSTO laboratories. Although many of these practical activities have global application, Australian nuclear research institutions, regulators and safeguards officials have actively helped countries in Australia's immediate region implement nuclear safeguards, export controls, and regulatory and safety measures, and thus worked in direct support of Australia's strategic interest in keeping nuclear proliferation far from its shores.

One of the aims of this book has been to demonstrate that while the tone and activism of Australian non-proliferation diplomacy has reflected and reinforced Australia's status as a non-proliferation champion, the country's contributions

9 For one of the better analyses of norms and their role in reinforcing non-proliferation decisions, see Maria Rost Rublee, *Nonproliferation Norms: Why States Choose Nuclear Restraint* (Athens and London: University of Georgia Press, 2009).

10 For discussion, see Thomas Wilkins, 'Australia: A Traditional Middle Power Faces the Asian Century', in Bruce Gilley and Andrew O'Neil, eds, *Middle Powers and the Rise of China* (Georgetown: Georgetown University Press, 2014), 149–70.

have been more nuanced and strategic than a focus on the efforts of Australian diplomats alone would suggest. Because Australia's main economic benefit from the civilian use of nuclear technology has relied on the spread of nuclear power to other countries, the dual-use risk has remained important in nuclear policy debates. Mastering the details of nuclear technology, as well as retaining the personnel with knowledge and experience, has contributed to bolstering non-proliferation and safeguards in other countries. However, this has always rested on a very narrow basis in terms of funding, personnel and expertise, which is becoming increasingly difficult as budget cuts to key agencies bite harder and investment in maintaining Australia's understanding of fuel cycle technologies has contracted over time. Australia's future role as a leader in nuclear non-proliferation will depend to a large extent on how successful it is in its own region in building firewalls against the spread of nuclear weapons. Speeches and political exhortations will have their place, but ultimately Australian nuclear leadership will rise or fall on the basis of how much governments are willing to invest in helping to proliferation proof the Asia-Pacific, a region where robust demand for nuclear power shows no sign of abating in the 21st century.

Bibliography

The Acronym Institute. 'Aftermath of US Senate Rejection of CTBT'. *Disarmament Diplomacy* 41 (November 1999). Available at http://www.acronym.org.uk/dd/dd41/41after.html.

Alagappa, Muthiah, ed. *Security in 21st Century Asia*. Stanford, CA: Stanford University Press, 2008.

Albinski, Henry S. 'Problems in Australian Foreign Policy – July to December 1977'. *Australian Journal of Politics and History* 24, no. 1 (April 1978): 1–19.

Albright, David and Corey Hinderstein. 'Unraveling the A.Q. Khan Network and Future Proliferation Networks'. *Washington Quarterly* 28, no. 2 (2005): 111–28.

Allard, Tom. 'Defence Push for Australian Missile Shield'. *Sydney Morning Herald*. 27 February 2003. Available at http://www.smh.com.au/articles/2003/02/26/1046064107709.html.

Allison, Graham T. *Essence of Decision: Explaining the Cuban Missile Crisis*. Boston: Little, Brown and Company, 1971.

Anderson, David. 'Nuclear Safeguards'. Research Paper prepared for the Senate Uranium Mining and Milling Committee, May 1997. Available at http://www.aph.gov.au/Parliamentary_Business/Committees/Senate/Former_Committees/uranium/report/c12.

Anon. 'Don't Close Door to A-Weapons: Minister'. *The Sydney Morning Herald*. 14 July 1969.

———. 'Scherger Wants Us to Have Nuclear Bombs'. *The Australian*. 8 October 1971.

———. 'NT Uranium Project "New Basic Industry"'. *The Canberra Times*. 10 September 1975.

———. 'Mary Kathleen Uranium: Export Permits Allowed'. *The Canberra Times*. 12 November 1976.

———. 'Australian Uranium: Anthony Supports Carter'. *The Economist*. 21 May 1977.

———. 'The Anti-Bomb-Spread Treaty is Still Worth Saving'. *The Economist*. 16 August 1980.

———. 'US Embassy Cables: Hillary Clinton Ponders US Relationship with its Chinese "Banker"'. *The Guardian*. 5 December 2010.

Arnett, Eric. 'The Comprehensive Nuclear Test Ban'. *SIPRI Yearbook 1995: Armaments, Disarmament and International Security*. Oxford: Oxford University Press, 1995.

Aron, Joan. *Licensed to Kill? The Nuclear Regulatory Commission and the Shoreham Power Plant*. Pittsburgh, PA: Pittsburgh University Press, 1998.

Atlas, Terry and James O'Shea. 'The U.S. to Test MX without Australian Aid'. *Chicago Tribune.* 7 February 1985.

Aulich, Chris and Mark Evans, eds. *The Rudd Government: Australian Commonwealth Administration, 2007–2010.* Canberra: ANU E-Press, 2010.

Australian Atomic Energy Commission. 'Jervis Bay Nuclear Power Project'. *Atomic Energy in Australia* 13, no. 2 (April 1970).

———. *Australian Atomic Energy Commission Annual Report: 1976–77.* Canberra: Australian Atomic Energy Commission, 1977.

Australian Broadcasting Corporation (ABC Online). 'Rudd Slams Indian Uranium Decision'. 15 August 2007. Available at http://abc.gov.au/news/stories/2007/08/15/2005697.htm.

———. 'Labor Pledges to Overturn India Uranium Deal'. 17 August 2007. Available at http://www.abc.net.au/news/stories/2007/08/17/2007382.htm.

———. 'Govt. Reverses Uranium Sale to India'. 15 January 2008. Available at http://www.abc.net.au/news/stories/2008/01/15/2139188.htm.

———. 'Rudd Heralds New Nuclear Disarmament Body'. 9 June 2008. Available at http://www.abc.net.au/news/2008-06-09/rudd-heralds-new-nuclear-disarmament-body/2465316.

Australian Broadcasting Corporation (*7:30 Report*). 'Kevin Rudd, TV Interview'. 30 April 2007. Available at http://www.alp.org/media/0407/tviloo301.php?mode=print.

Australian Department of Defence. *The Defence of Australia.* Canberra: Australian Government Publishing Service, March 1987.

———. *Australia's Strategic Planning in the 1990s.* Canberra: Australian Government Publishing Service, 1992.

———. *Strategic Review 1993.* Canberra: Australian Government Publishing Service, 1993.

———. *Defending Australia: Defence White Paper 1994.* Canberra: Australian Government Publishing Service, 1994.

———. *Australia's Strategic Policy.* Canberra: Commonwealth of Australia, December 1997.

———. *Australia's National Security: A Defence Update 2003.* Canberra: Commonwealth of Australia, 2003.

———. *Defending Australia in the Asia Pacific Century: Force 2030.* Canberra: Commonwealth of Australia, 2009.

———. 'Draft Submission to the United States Nuclear Posture Review 2009'. Available at http://www.defence.gov.au/foi/Decisions/DisclosureLog.asp.

———. *Defence White Paper 2013.* Canberra: Commonwealth of Australia, 2013.

Australian Department of Foreign Affairs. 'Draft Treaty on the Non-Proliferation of Nuclear Weapons'. *Current Notes on International Affairs* (Statements) 39, no. 5 (May 1968): 206–10.

———. 'Ratification of the Nuclear Non-Proliferation Treaty'. *Australian Foreign Affairs Record* (Statements) 44, no. 2 (February 1973).

———. 'Agreement between Australia and the International Atomic Energy Agency for the Application of Safeguards in Connection with the Treaty on the Non-Proliferation of Nuclear Weapons: Entry into Force, 10 July 1974'. *Treaty Series 1974*, 16. Canberra: Australian Government Publishing Service, 1974.

———. 'Swift Action Needed to Curb Proliferation: Text of a Statement Made on 8 May 1975 by Mr R.W. Furlonger, Leader of the Australian Delegation to the Review Conference of the Parties to the Treaty on the Non-Proliferation of Nuclear Weapons'. *Australian Foreign Affairs Record* (Statements) 46, no. 5 (May 1975): 280–81.

———. 'Foreign Minister Addresses UN General Assembly'. *Backgrounder* (Statements) 59 (1 October 1976).

———. *Uranium, the Joint Facilities, Disarmament and Peace.* Canberra: Australian Government Publishing Service, 1984.

———. *Third Review Conference of the Parties to the Treaty on the Non-Proliferation of Nuclear Weapons, Geneva, 27 August to 21 September 1985: Report of the Australian Delegation.* Canberra: Australian Government Publishing Service, 1985.

———. *Fourth Review Conference of the Parties to the Treaty on the Non-Proliferation of Nuclear Weapons, Geneva, 20 August to 15 September 1990: Report of the Australian Delegation.* Canberra: Australian Government Publishing Service, 1990.

Australian Department of Foreign Affairs and Trade. 'Nuclear Non-proliferation: Fourth Review Conference'. *Minister for Foreign Affairs and Trade News Release*, no. M158 (17 September 1990).

———. 'Nuclear Suppliers Group to Strengthen Information Exchange, Conditions of Supply'. *Peace and Disarmament News* (March 1992).

———. *Department of Foreign Affairs and Trade Annual Report: 1995–1996.* Canberra: Australian Government Publishing Service, 1996.

———. *Report of the Canberra Commission on the Elimination of Nuclear Weapons: August 1996.* Canberra: Australian Government Publishing Service, 1996.

———. *Agreement Between the Government of Australia and the Government of the People's Republic of China for Cooperation in the Peaceful Uses of Nuclear Energy.* Canberra, 3 April 2006. Available at http://www.dfat.gov.au/geo/china/treaties/nuclear_energy.html.

Australian Department of Prime Minister and Cabinet. *Securing Australia's Energy Future.* Canberra: Commonwealth of Australia, 2004.

———. *Uranium Mining, Processing and Nuclear Energy – Opportunities for Australia?* Canberra: Commonwealth of Australia, 2006.

Australian Labor Party (ALP). *Australian Labor Party Platform, Constitution and Rules as Approved by the 36th National Conference.* Canberra: National Secretariat, 1984.

Australian Nuclear Science and Technology Organisation. *ANSTO Annual Report 1993–1994*. Lucas Heights: ANSTO, 1994.

Australian Safeguards and Non-Proliferation Office. *Australian Safeguards and Non-Proliferation Office Annual Report: 1998–1999*. Canberra: Australian Government Publishing Service, 1999.

———. *Annual Report 2011–2012*. Canberra: Australian Government Publishing Service, 2012.

Australian Safeguards Office. *Annual Report 1993–1994*. Canberra: Australian Government Publishing Service, 1994.

———. *Annual Report of the Director of Safeguards 1994–1995*. Canberra: Australian Government Publishing Service, 1995.

Auton, Luke. 'Opaque Proliferation: The Historiography of Australia's Cold War Nuclear Weapons Option'. *History Compass* 11, no. 8 (2013): 561–72.

Bachelard, M. 'Introducing Paul Keating's First XV for the No Nukes Cup'. *The Canberra Times*. 27 November 1995.

Baker, Steven J. 'Monopoly or Cartel?' *Foreign Policy* 23 (Summer 1976): 202–20.

Ball, Desmond. *A Suitable Piece of Real Estate*. Sydney: Hale & Iremonger, 1980.

———. 'Australia and Nuclear Policy'. In Desmond Ball, ed., *Strategy & Defence*. Sydney: George Allen & Unwin, 1982. 320–42.

———, ed. *Strategy & Defence*. Sydney: George Allen & Unwin, 1982.

———. 'The Strategic Essence'. *Australian Journal of International Affairs* 55, no. 20 (2001): 235–48.

Barkenbus, Jack. 'Whither the Treaty?' *Bulletin of the Atomic Scientists* 36, no. 4 (April 1980): 37–9.

Barker, Geoffrey. 'PM on Uranium: Never Say Never'. *The Weekend Australian Financial Review*. 18–19 March 2006.

Beaton, Leonard. 'Nuclear Fuel-For-All'. *Foreign Affairs* 45, no. 4 (1967): 662–9.

Beaufre, André. *An Introduction to Strategy*, trans. Major-General R.H. Barry. New York: Frederick A. Praeger, 1965.

Beeson, Mark. 'Australia and Asia: The Years of Living Aimlessly'. *Southeast Asian Affairs* (2001): 33–44.

Belcher, Emma. 'A Tighter Net: Strengthening the Proliferation Security Initiative'. *Lowy Institute Policy Brief*, August 2009. Available at http://www.lowyinstitute.org/publications/tighter-net-strengthening-proliferation-security-initiative.

Bell, Coral. 'Mr. Fraser and Australian Foreign Policy'. *The World Today* 35, no. 10 (1979): 414–20.

———. 'Australia in a World of Powers'. In P.J. Boyce and J.R. Angel, eds, *Independence and Alliance: Australia in World Affairs 1975–1980*. Sydney: George Allen & Unwin, 1983.

Bell, Roger. 'Reassessed: Australia's Relationship with the United States'. In James Cotton and John Ravenhill, eds, *Seeking Asia Engagement: Australia in World Affairs, 1991–95*. Melbourne: Oxford University Press, 1997.

Bellamy, C. 'Big Five Nuclear Powers Braced for Move to Ban Bomb'. *The Independent.* 4 January 1994.

Bellany, Ian. 'How Nuclear Should Australia Go?' *The Bulletin.* 23 August 1969. 43–5.

———. *Australia in the Nuclear Age: National Defence and National Development.* Sydney: Sydney University Press, 1972.

Binnie, Anna Eugenia. *From Atomic Energy to Nuclear Science: A History of the Australian Atomic Energy Commission.* Doctor of Philosophy Thesis, Macquarie University, 2003.

Blainey, Geoffrey. *The Rush that Never Ended: A History of Australian Mining*, 3rd edition. Melbourne: Melbourne University Press, 1978.

Bodansky, David. *Nuclear Energy: Principles, Practices, and Prospects.* New York: Springer, 2004.

Bonner, Raymond. 'Call to Enrich Uranium Stirs Debate in Australia'. *New York Times.* 2 August 2006. Available at http://www.nytimes.com/2006/08/02/world/asia/02australia.html?pagewanted=print&_r=0.

Boyce, P.J. and J.R. Angel, eds. *Independence and Alliance: Australia in World Affairs 1975–1980.* Sydney: George Allen & Unwin, 1983.

Bracken, Paul. 'The Second Nuclear Age'. *Foreign Affairs* 79, no. 1 (January/February 2000).

Bratt, Duane. *Canada, the Provinces and the Global Nuclear Revival.* Montreal: McGill-Queen's University Press, 2012.

Braun, Chaim. 'The Nuclear Energy Market and the Nonproliferation Regime'. *The Nonproliferation Review* 13, no. 3 (2006): 627–44.

——— and Christopher Chyba. 'Proliferation Rings: New Challenges to the Nuclear Non-Proliferation Regime'. *International Security* 29, no. 2 (2004): 5–49.

Brodie, Bernard. 'Implications for Military Policy'. In Bernard Brodie, ed., *The Absolute Weapon.* New York: Harcourt, Brace and Company, 1946.

Broinowski, Richard. *Fact or Fission? The Truth about Australia's Nuclear Ambitions.* Melbourne: Scribe, 2003.

Brown, Colin. 'Problems in Australian Foreign Policy, January–June 1996'. *Australian Journal of Politics and History* 42, no. 3 (1996): 331–44.

Brown, Nanette C. *The Strategic Defense Initiative and European Security: A Conference Report.* Santa Monica, CA: RAND, 1986.

Bull, Hedley. 'Australia and the Nuclear Problem: Some Concluding Comments'. In Robert O'Neill, ed., *The Strategic Nuclear Balance: An Australian Perspective.* Canberra: Australian National University Press, 1975.

Burns, A.L. 'Australia and the Nuclear Balance'. In Harry Gelber, ed., *Problems of Australian Defence.* Melbourne: Oxford University Press, 1970.

Byers, Michael. 'Policing the High Seas: The Proliferation Security Initiative'. *The American Journal of International Law* 98, no. 3 (2004): 526–45.

Campbell, D. 'Utopian Gesture Will Not Disarm the Warheads'. *The Australian*. 22 January 1996.

Carlson, John. 'Safeguards and Non-Proliferation: Current Challenges and the Implications for Australia'. Paper presented to the 2005 Conference of the Australian Nuclear Association, Sydney. 10 November 2005.

———. *SAGSI: Its Role and Contribution to Safeguards Development*. Barton: Australian Safeguards and Non-Proliferation Office, 2007. Available at http://www.dfat.gov.au/asno/publications/SAGSI_role_contribution_safeguards_dev.pdf.

———. 'IAEA Safeguards Additional Protocol'. 20 January 2009. Available at https://www.dfat.gov.au/asno/publications/iaea-safeguards-additional-protocol.html.

———. 'The "Prague Agenda" and Nuclear Energy'. Australian Embassy, Washington, DC, 13 November 2012. Available at http://www.lowyinstitute.org/publications/prague-agenda-and-nuclear-energy.

———. 'The Potential for a Regional Uranium Enrichment Centre in Australia'. *Security Challenges* 9, no. 2 (2013): 1–9.

———. 'Is the Abbott Government Abandoning Australia's Nuclear Safeguards Standards for India?' *The Interpreter*. 1 October 2014. Available at http://www.lowyinterpreter.org/post/2014/10/01/Is-the-Abbott-Government-abandoning-Australias-nuclear-safeguards-standards-for-India.aspx.

———, J. Bardsley and J. Hill. 'Trials in Australia related to the assurance of the absence of significant undeclared nuclear activities (IAEA-SM-333/132P)'. In International Atomic Energy Agency, ed., *International Nuclear Safeguards 1994: Vision for the Future*. Vienna: IAEA, 1994. 290–93.

Carnegie Endowment for International Peace. 'About the Carnegie Nuclear Policy Program'. Available at http://carnegieendowment.org/programs/npp/?fa=about.

Carr, Andrew. *Winning the Peace: Australia's Campaign to Change the Asia Pacific* (Melbourne: Melbourne University Press, 2014).

——— and Chris Roberts. 'Foreign Policy'. In Chris Aulich and Mark Evans, eds, *The Rudd Government: Australian Commonwealth Administration, 2007–2010*. Canberra: ANU E-Press, 2010.

Carr, Bob. 'Australia-UAE Nuclear Cooperation Agreement'. Media Release – Minister for Foreign Affairs, 1 August 2012. Available at http://foreignminister.gov.au/releases/2012/bc_mr_120801a.html.

Carr, Michael. 'Australia and the Nuclear Question: A Survey of Government Attitudes, 1945 to 1975'. Unpublished Master of Arts Thesis, University of New South Wales, 1979.

Carranza, Mario E. 'Can the NPT Survive? The Theory and Practice of US Nuclear Non-Proliferation Policy after September 11'. *Contemporary Security Policy* 27, no. 3 (2006): 489–525.

Carroll, J. 'Ireland Breaks EU Ranks and Supports Move to Ban Nuclear Weapons'. *The Irish Times*. 10 June 1995.

Carter, Ashton B. 'America's New Strategic Partner?' *Foreign Affairs* 85, no. 4 (2006): 33–44.

Catley, Robert. 'Australia and the Great Powers'. *Australian Outlook* 37, no. 3 (1983): 143–9.

Cawte, Alice. *Atomic Australia 1944–1990*. Sydney: UNSW Press, 1992.

Center for Strategic and International Studies Nuclear Issues Project. 'Nuclear Issues Project'. Available at http://csis.org/program/nuclear-policy-news.

Chernus, Ira. *Apocalypse Management: Eisenhower and the Discourse of National Insecurity*. Stanford, CA: Stanford University Press, 2008.

Cirincione, Joseph. 'Why the Right Lost the Missile Defense Debate'. *Foreign Policy* 106 (Spring 1997): 38–55.

———. 'The Asian Nuclear Reaction Chain'. *Foreign Policy* (Spring 2000): 120–36.

———. 'The US's Nuclear Cave In'. *Asia Times*. 4 March 2006. Available at http://www.atimes.com/atimes/South_Asia/HC04Df03.html.

Clarke, Michael. 'Refashioning Australia's Nuclear Bargain? The Challenges of Changing Strategic, Regime and Market Environments'. *The Nonproliferation Review* 15, no. 2 (2008): 311–34.

———. 'Australia, India and the Uranium Question'. *Australian Journal of Political Science* 46, no. 3 (2011): 489–502.

———. 'The Fraser Government's "Uranium Decision" and the Foundations of Australian Non-Proliferation Policy: A Reappraisal'. *Australian Journal of Politics and History* 58, no. 2 (2012): 221–35.

———, Stephan Frühling and Andrew O'Neil, eds. *Australia's Uranium Trade: The Domestic and Foreign Policy Challenges of a Contentious Export*. Farnham: Ashgate, 2011.

Clary, Christopher. 'The AQ Khan Network and the Limits of the Non-Proliferation Regime'. *Disarmament Forum* 4 (2004): 33–42.

Cockburn, Stewart and David Ellyard. *Oliphant: The Life and Times of Sir Mark Oliphant*. Adelaide: Axiom Books, 1981.

Committee of Review of the Australian Atomic Energy Commission. *Review of the Australian Atomic Energy Commission*. Canberra: Australian Government Publishing Service, 1986.

Commonwealth of Australia. *Atomic Energy Act 1953–1973*. Available at http://www.comlaw.gov.au/Details/C2004C01926.

———. 'Threats to Australia: Their Nature and Probability'. *Joint Committee on Foreign Affairs and Defence*. Canberra: Australian Government Publishing Service, 1981.

———. 'Nuclear Non-Proliferation (Safeguards) (Consequential Amendments) Bill 1987: Explanatory Memorandum'. Available at http://www.austlii.edu.au/au/legis/cth/bill_em/nnab1987614/memo_0.html.

———. 'The 1998 Indian and Pakistani Nuclear Tests'. *Report of the Senate Foreign Affairs, Defence and Trade References Committee.* Canberra: Commonwealth of Australia, June 1999. Available at http://www.aph.gov.au/~/media/wopapub/senate/committee/fadt_ctte/completed_inquiries/1999_02/nuclear/report/report_pdf.ashx.

———. 'Proposed Replacement Nuclear Research Reactor, Lucas Heights, NSW'. *Parliamentary Standing Committee on Public Works.* Canberra: Commonwealth of Australia, 12 August 1999.

———. 'Inquiry into Nuclear Non-proliferation and Disarmament'. *Parliamentary Joint Standing Committee on Treaties, Report no. 106.* Canberra: Commonwealth of Australia, September 2009.

Commonwealth of Australia Parliamentary Debates (Hansard) (*C.P.D.*). House of Representatives (*Representatives*), Senate (*Senate*).

Conkey, H. 'N-Free Zone Plan "So Defective"'. *The Canberra Times.* 29 July 1985.

Cooper, David A. *Competing Western Strategies Against the Proliferation of Weapons of Mass Destruction.* Westport, CT: Praeger, 2002.

———. 'Challenging Contemporary Notions of Middle Power Influence: Implications of the Proliferation Security Initiative for "Middle Power Theory"'. *Foreign Policy Analysis* 7, no. 3 (2011): 317–36.

Coorey, Phillip. 'Gillard's Push for Uranium Sales to India'. *Sydney Morning Herald.* 15 November 2011.

Cotton, James and David Lee, eds. *Australia and the United Nations.* Canberra: Commonwealth of Australia, 2012.

Cotton, James and John Ravenhill, eds. *Seeking Asia Engagement: Australia in World Affairs, 1991–95.* Melbourne: Oxford University Press, 1997.

Counterproliferation Program Review Committee. *Report on Activities and Programs for Countering Proliferation and NBC Terrorism, Executive Summary.* Washington, DC: Government Printing Office, October 2001. Available at http://www.nti.org/media/pdfs/40_5.pdf?=1318280983.

Cross, Frank B. and Cyril Smith. 'The Reagan Administration's Nonproliferation Nonpolicy'. *Catholic University Law Review* 33 (1983/84): 633–65.

Curran, Enda. 'Australia Eyes UAE Uranium Talks'. *Wall Street Journal.* 9 March 2011.

D'Apulget, Blanche. *Hawke: The Prime Minister.* Melbourne: Melbourne University Press, 2011.

Defence Committee. *Strategic Basis of Australian Defence Policy 1956.* Canberra: Department of Defence, 11 October 1956.

———. *Strategic Basis of Australian Defence Policy 1971.* Canberra: Department of Defence, 5 March 1971.

———. *Strategic Basis of Australian Defence Policy 1973.* Canberra: Department of Defence, 1 June 1973.

———. *Strategic Basis of Australian Defence Policy 1975.* Canberra: Department of Defence, 3 October 1975.

———. *Australian Strategic Analysis and Defence Policy Objectives 1976*. Canberra: Department of Defence, 1976.

———. *Strategic Basis of Australian Defence Policy 1983*. Canberra: Department of Defence, 15 August 1983.

Department of Safeguards. *Development and Implementation Support Programme for Nuclear Verification 2012–2013*, STR-371. Vienna: IAEA, 2011.

Deutch, John, Arnold Canter, Ernest Moniz and Daniel Poneman. 'Making the World Safe for Nuclear Energy'. *Survival* 46 (Winter 2004/05): 65–80.

Dewes, K. and R. Green. 'The World Court Project: How a Citizen Network Can Influence the United Nations'. *Pacifica Review* 7, no. 2 (October/November 1995): 17–37.

Dewitt, D., ed. *Nuclear Non-Proliferation and Global Security*. New York: St. Martin's Press, 1987.

Dibb, Paul. *Review of Australia's Defence Capabilities*. Canberra: Australian Government Publishing Service, 1986.

Division of Physical and Chemical Sciences. *Research Reactors: Purpose and Future*. Vienna: International Atomic Energy Agency, 2010.

Dobbell, Graeme. 'Des Ball: An Australian Original'. *Lowy Interpreter Blog*. 17 December 2013. Available at http://www.lowyinterpreter.org/post/2012/12/17/Des-Ball-An-Australian-original.aspx.

Dodge, Michaela and Baker Spring. 'Bait and Switch on Nuclear Modernization Must Stop'. *Backgrounder* 2755. Washington, DC: Heritage Foundation, 2013.

Donnelly, Warren H. 'INFCE and US Non-Proliferation Policy'. *Energy Policy* 8, no. 4 (1980): 335–8.

———. 'Nonproliferation Policy of the United States in the 1980s'. *SAIS Review* 7, no. 2 (1987): 167–71.

Dorling, Philip. 'ALP Nuclear Backflip Linked to US Defence'. *The Sydney Morning Herald*. 2 October 2013.

Dossani, Raaq and Henry S. Rowen, eds. *Prospects for Peace in South Asia*. Stanford, CA: Stanford University Press, 2005.

Downer, Alexander. 'Australia and Asia: Taking the Long View'. Address by Mr Alexander Downer, MP, Minister for Foreign Affairs, to the Foreign Correspondents' Association. Sydney, 11 April 1996. Available at http://www.foreignminister.gov.au/speeches/1996/asia-long.html.

———. 'Security through Cooperation'. Address by Mr Alexander Downer, MP, Minister for Foreign Affairs, to the IISS/SDSC Conference, 'The New Security Agenda in the Asia-Pacific Region'. Canberra, 2 May 1996. Available at http://www.foreignminister.gov.au/1996/regsec5.html.

———. 'Australia and the United States: A Vital Friendship'. Speech by Mr Alexander Downer, MP, Minister for Foreign Affairs, to the Australian Centre for American Studies. Sydney, 29 May 1996. Available at http://www.dfat.gov.au/media/speeches/foreign/1996/aust_uni.html.

———. 'Chinese Nuclear Test'. Department of Foreign Affairs and Trade Media Release, 8 June 1996. Available at http://foreignminister.gov.au/releases/1996/fa46.html.

———. 'Australia Signs the Comprehensive Test Ban Treaty (CTBT)'. Department of Foreign Affairs and Trade Media Release, 25 September 1996. Available at http://www.foreignminister.gov.au/1996/fa99.html.

———. 'The United Nations for the 21st Century'. Address to the 51st General Assembly of the United Nations by Australia's Minister for Foreign Affairs. New York, 30 September 1996. Available at http://www.dfat.gov.au/cc/947_3.html.

———. 'Address by the Hon. Alexander Downer, MP, Minister for Foreign Affairs, to the Conference on Disarmament'. Geneva, 30 January 1997. Available at http://www.dfat.gov.au/media/speeches/foreign/1997/geneva_c.html.

———. 'Two Swords for the Beating'. Address by the Hon. Alexander Downer, MP, Minister for Foreign Affairs, to the Conference on Disarmament. Geneva, 3 February 1998. Available at http://www.dfat.gov.au/media/speeches/foreign/1998/swords3feb98.html.

———. 'The NPT – Future Challenges and Outlook for the 2000 Review Conference'. Keynote Speech by the Hon. Alexander Downer, MP, Minister for Foreign Affairs, to the Conference on Nuclear Arms Control: Australian and Regional Perspectives Towards 2000. University of Queensland, 29 October 1999. Available at http://www.dfat.gov.au/media/speeches/foreign/1999/991029_npt.html.

———. 'Speech by the Minister for Foreign Affairs'. NPT Review Conference. New York, 25 April 2000. Available at http://www.dfat.gov.au/media/speeches/foreign/2000/000425_npt.html.

———. 'Weapons of Mass Destruction: The Greatest Threat to International Security'. Speech by the Hon. Alexander Downer, MP, Minister for Foreign Affairs, to the Proliferation Security Initiative. Brisbane, 9 July 2003. Available at http://www.foreignminister.gov.au/speeches/2003/030709_wmd.html.

———. 'The Threat of Proliferation: Global Resolve and Australian Action'. Speech to the Lowy Institute by the Hon. Alexander Downer, MP, Minister for Foreign Affairs. Sydney, 23 February 2004. Available at http://www.foreignminister.gov.au/speeches/2004/04223_lowy.html.

———. 'Statement to the Seventh Nuclear Non-Proliferation Treaty Review Conference by the Australian Minister for Foreign Affairs, the Hon. Alexander Downer, MP'. New York, 2 May 2005. Available at http://www.foreignminister.gov.au/speeches/2005/050502_7th_nuclear_conference.html.

———. 'Transcript of the Foreign Minister the Hon. Alexander Downer, MP, Doorstop Interview'. Hyatt Hotel, Adelaide, 3 March 2006. Available at http://www.foreignminister.gov.au/transcripts/2006/060303_ds.html.

———. 'Joint Press Conference between Foreign Minister Alexander Downer and US Secretary of State, Condoleezza Rice'. 16 March 2006. Available at http://www.foreignminister.gov.au/transcripts/2006/060316_joint_press_conference.html.

———. 'The Impact of Energy Security on Australia's International Relations'. Speech to the Australian Institute of International Affairs. 11 October 2006. Available at http://www.foreignminister.gov.au/speeches/2006/061011_es.html.

———. 'Speech and Question and Answer Session'. Energy Environment and Air Quality Policy Forum. Los Angeles, 13 January 2007. Available at http://www.foreignminister.gov.au/transcripts/2007/070113_eeaqpf.html.

——— and Peter McGuaran. 'Boost for Australia's Nuclear Watchdog Role'. Joint Media Release. Available at http://www.foreignminister.gov.au/releases/2002/fa065j_02.html.

——— and Ian McLachlan. 'Australia-United States: A Strategic Partnership for the Twenty-First Century: Sydney Statement'. Joint Security Declaration. Sydney, 27 July 1996. Available at https://www.dfat.gov.au/geo/us/ausmin/sydney_statement.html.

Editor. *Arms Control Reporter* 4, no. 4 (April 1985).

———. *Arms Control Reporter* 13, no. 4 (April 1994).

———. 'Australia Puts Forward Draft Treaty Elements'. *Disarmament: Newsletter of the UN Centre for Disarmament Affairs* 12, no. 2 (1994).

———. 'The ICJ and Nuclear Arms'. *The Canberra Times*. 7 February 1995.

———. 'CTB Negotiations: Geneva Update'. *Nuclear Proliferation News*, no. 29 (11 July 1995).

———. 'Use of Force and War: Weapons of Mass Destruction – Prevention of Proliferation – New Australian Legislation'. *Australian Year Book on International Law* 26 (1995): 549–72.

Elert, Glenn. 'Albert Einstein's Letters to President Franklin Delano Roosevelt'. Available at http://hypertextbook.com/eworld/einstein.shtml#first.

Emery, Brigadier General. 'Statement of Brigadier General Emery, USAF, Deputy for Theatre and Air Missile Defense, Ballistic Missile Defense Organization, to the House National Security Committee, Subcommittees on Research & Development and Procurement'. Washington, DC, 19 March 1997. Available at http://www.mda.mil/global/documents/pdf/ps_emery.pdf.

Encel, S. and A. McKnight. 'Bombs, Power Stations, and Proliferation'. *Australian Quarterly* 42, no. 1 (March 1970): 17–21.

Evans, Chris. 'Labor and Uranium – An Evolution'. Speech by Senator Chris Evans, Leader of the Opposition in the Senate, Shadow Minister for National Development, Resources and Energy, to the Labor Business Roundtable. Parmelia Hilton, Perth, 23 March 2007. Available at http://www.alp.org.au.

Evans, Gareth. *Australia's Regional Security*. Canberra: Department of Foreign Affairs and Trade, 1989.

Ferguson, Gregor. 'Australia, U.S. Discuss SBIRS Collaboration Effort'. *Defense News* 14, no. 42 (25 October 1999): 10.

Findlay, Trevor. 'A Comprehensive Nuclear Test Ban: Post-Cold War Prospects'. *Peace Research Centre Working Paper*, no. 122. Canberra: Australian National University, September 1992.

———. 'Explaining Australasian Angst: Australia, New Zealand and French Nuclear Testing'. *Security Dialogue* 26, no. 4 (1995): 373–81.

———. 'The Nuclear Energy Revival and Global Governance'. In Michael Clarke, Stephan Frühling and Andrew O'Neil, eds, *Australia's Uranium Trade: The Domestic and Foreign Policy Challenges of a Contentious Export*. Farnham: Ashgate, 2011.

Finnemore, Martha and Kathryn Sikkink. 'International Norm Dynamics and Political Change'. *International Organization* 52, no. 4 (1998): 887–917.

Fischer, David A. 'Viewpoint: New Directions and Tools for Strengthening IAEA Safeguards'. *The Nonproliferation Review* (Winter 1996): 69–76.

——— and Harald Müller. 'The Fourth Review Conference of the Non-Proliferation Treaty'. In Stockholm International Peace Research Institute, ed., *World Armaments and Disarmament: SIPRI Yearbook 1991*. Oxford: Oxford University Press, 1991.

Flitton, Daniel. 'Abbott's "Anglosphere" Looks to the Past'. *Sydney Morning Herald*. 27 April 2010.

——— and Ben Nicholson. 'Putin's Ultimatum to Rudd over Uranium Deal'. *The Age*. 2 September 2008.

Ford, Gerald R. 'Statement on Nuclear Policy'. 28 October 1976. Online by Gerhard Peters and John T. Woolley. *The American Presidency Project*. Available at http://www.presidency.ucsb.edu/ws/?pid=6561.

Fox, Russell W. *Ranger Uranium Environmental Inquiry: First Report*. Canberra: Australian Government Publishing Service, 1976.

Franko, Lawrence G. 'US Regulation of the Spread of Nuclear Technologies through Supplier Power: Lever or Boomerang?' *Law and Policy in International Business* 10, no. 4 (1978): 1181–204.

Fraser, Malcolm. 'National Objectives – Social, Economic, and Political Goals'. *Australian Quarterly* 47, no. 1 (1975): 24–35.

Freedman, Lawrence. *The Evolution of Nuclear Strategy*, 3rd edition. London: Palgrave Macmillan, 2003.

———. *Deterrence*. Cambridge: Polity Press, 2004.

———. 'Disarmament and Other Nuclear Norms'. *Washington Quarterly* 36, no. 2 (2013): 93–108.

Frühling, Stephan. 'Ballistic Missile Defence for Australia: Policies, Requirements and Options'. *Canberra Papers on Strategy and Defence*, no. 151. Canberra: Strategic and Defence Studies Centre, ANU, September 2003.

———. 'Never Say Never: Considerations about the Possibilities of Australia Acquiring Nuclear Weapons'. *Asian Security* 6, no. 2 (2010): 146–69.

Fuhrmann, Matthew. 'Spreading Temptation: Proliferation and Peaceful Nuclear Cooperation'. *International Security* 34, no. 1 (2009): 7–41.

———. *Atomic Assistance: How 'Atoms for Peace' Programs Cause Nuclear Insecurity*. Ithaca, NY: Cornell University Press, 2012.

Garnaut, John. 'China Breaches Spur Nuclear Fallout'. *The Age*. 27 August 2007.

Garthoff, R.L. *Detente and Confrontation: American-Soviet Relations from Nixon to Reagan*. Washington, DC: The Brookings Institution, 1985.

Garver, J. *Face-Off: China, the United States, and Taiwan's Democratization*. Seattle: University of Washington Press, 1997.

Gavin, Francis J. 'The Myth of Flexible Response: United States Strategy in Europe during the 1960s'. *International History Review* 23, no. 3 (2001): 847–75.

George Washington University Elliott School of International Affairs. 'Nuclear Policy Talks'. Available at http://elliott.gwu.edu/news/events/nuclear-policy-talks.cfm.

Gibson, R. 'Nuclear Experts Told to Act Now'. *The Age*. 23 January 1996.

Gillard, Julia. 'An Opportunity for Sound and Fury to Signify Something'. *The Age*. 15 November 2011. Available at http://www.theage.com.au/opinion/an-opportunity-for-sound-and-fury-signifying-something-20111114-1nfi3.html.

Girling, J.L.S. 'Australia and Southeast Asia in the Global Balance: A Critique of the "Fraser Doctrine"'. *Australian Outlook* 31, no. 1 (1977): 3–15.

Glenn, William G. 'Commercial Liberties and Nuclear Anxieties: The US-German Feud over Brazil, 1975–7'. *International History Review* 34, no. 3 (2012): 449–74.

GlobalSecurity.org Website. 'DUNDEE'. Available at http://www.globalsecurity.org/space/systems/dundee.html.

Goldblat, J. *Twenty Years of the Non-Proliferation Treaty: Implementation and Prospects*. Oslo: International Peace Research Institute, 1990.

——— and S. Lodgaard. 'The South Pacific Nuclear Free Zone: Variation on a Latin American Theme'. *New Zealand International Review* 11, no. 3 (May/June 1986): 14–15.

Gordon, Sandy. 'Implications of the Sale of Australian Uranium to India'. *Working Paper* no. 410. Canberra: Strategic and Defence Studies Centre, September 2008.

Graham, Thomas. 'The Duration of the Nuclear Non-Proliferation Treaty: Sudden Death or New Lease on Life?' *Virginia Journal of International Law* 29, no. 3 (1989): 661–77.

Grattan, M. 'PM Picks World Anti-Nuke Team'. *The Age*. 27 November 1995.

Greenless, Don and Richard MacGregor. 'Downer Warns China over War Games'. *The Australian*. 13 March 1996.

Greenpeace. *Greenpeace, Secrets, Lies and Uranium Enrichment: The Classified SILEX Project at Lucas Heights*. Sydney: Greenpeace, November 2004. Available at http://www.mapw.org.au/download/secrets-lies-and-uranium-enrichment-2004-greenpeace-report-silex.

Greenwood, Ted and Robert Haffa. 'Supply-Side Non-Proliferation'. *Foreign Policy*, no. 42 (Spring 1981): 125–40.

Gummett, Philip. 'From NPT to INFCE: Developments in Thinking about Nuclear Proliferation'. *International Affairs* 57, no. 4 (1981): 549–67.

Gurry, Meg. 'Identifying Australia's "Region": From Evatt to Evans'. *Australian Journal of International Affairs* 49, no. 1 (1995): 17–31.

Gyngell, Allan and Michael Wesley. *Making Australian Foreign Policy*, 2nd edition. Cambridge: Cambridge University Press, 2007.

Hansen, Chuck. *U.S. Nuclear Weapons: The Secret History*. Arlington, TX: Aerofax, 1988.

Hanson, Marianne. 'The Advocacy States: Their Normative Role before and after the U.S. Call for Nuclear Zero'. *The Nonproliferation Review* 17, no. 1 (2010): 71–93.

——— and Carl Ungerer. 'The Canberra Commission: Paths Followed, Paths Ahead'. *Australian Journal of International Affairs* 53, no. 1 (1999): 5–17.

Hardy, Clarence C. *Atomic Rise and Fall: The Australian Atomic Energy Commission 1953–1987*. Peakhurst, NSW: Glen Haven, 1999.

Harries, Owen. 'Punching above Our Weight?' Boyer Lecture, 21 December 2003. Available at http://www.abc.net.au/radionational/programs/boyerlectures/lecture-6-punching-above-our-weight/3459850.

Haupt, R. 'Soviets Match Arms Cuts'. *The Age*. 7 October 1991.

Hawke, Robert. 'Speech by the Australian Prime Minister to the Washington Press Club'. Washington Press Club. Washington, DC, 15 June 1983.

Hayden, Bill. *Uranium, the Joint Facilities, Disarmament and Peace*. Canberra: Australian Government Publishing Service, 1984.

———. *Hayden: An Autobiography*. Sydney: Harper Collins Publishers, 1996.

Hayden, Sheila. 'Hawke Sees Approval of Uranium Developments As Step Towards Modifying ALP U308 Policy'. *Nuclear Fuel* 8, no. 24 (21 November 1983): 2–3.

Helmreich, Jonathan. *Gathering Rare Ores: The Diplomacy of Uranium Acquisition, 1943–1954*. Princeton, NJ: Princeton University Press, 1986.

Hersh, Seymour. 'On the Nuclear Edge'. *The New Yorker*. 29 March 1993.

Hewlett, Richard and Jack Holl. *Atoms for Peace and War, 1952–1961: Eisenhower and the Atomic Energy Commission*. Berkeley: University of California Press, 1989.

Holland, Ian. 'Waste Not Want Not? Australia and the Politics of High-level Nuclear Waste'. *Australian Journal of Political Science* 37, no. 2 (2002): 283–301.

Holst, J.J. 'Small Powers in a Nuclear World'. In A. Schou and A.O. Brundtland, eds, *Small States in International Relations*. Stockholm: Almqvist and Wiksell, 1971.

Hooper, Richard. 'The IAEA's Additional Protocol'. *Disarmament Forum* 3 (1999): 7–16.

———. 'The Changing Nature of Safeguards'. *IAEA Bulletin* (June 2003): 7–11.

Howard, John. 'Transcript of the Prime Minister the Hon. John Howard, MP, Address to the National Press Club'. The Great Hall Parliament House, Canberra, 14 March 2003. Available at http://www.pm.gov.au/news/speeches/speech/74.html.

———. 'Australia in the World'. Transcript of the Prime Minister the Hon. John Howard, MP, address to the Lowy Institute for International Policy. Westin Hotel, Sydney, 31 March 2005. Available at http://www.pm.gov.au/news/speeches/speech1290.html.

———. 'Transcript of the Prime Minister the Hon. John Howard, MP, Joint Press Conference, with the President of the United States of America George W. Bush'. The White House, Washington, DC, 19 July 2005. Available at http://www.pm.gov.au/news/interviews/Interview1472.html.

———. 'Transcript of the Prime Minister the Hon. John Howard, MP, Doorstop Interview'. Taj Mahal Hotel, Delhi, 5 March 2006. Available at http://www.pm.gov.au/news/interviews/Interview1806.html.

Hubbard, Christopher. 'From Ambivalence to Influence: Australia and the Negotiation of the 1968 Nuclear Non-Proliferation Treaty'. *Australian Journal of Politics and History* 50, no. 4 (2004): 526–43.

Huisken, Ron. 'The Proliferation Security Initiative: Coming in from the Cold'. *APSNet Policy Forum*, 20 April 2006. Available at http://nautilus.org/apsnet/0613a-huisken-html/.

Hultman, Nathan and Jonathan Koomey. 'Three Mile Island: The Driver of US Nuclear Power's Decline?' *Bulletin of the Atomic Scientists* 69, no. 3 (2013): 63–70.

Hymans, Jacques. 'Isotopes and Identity: Australia and the Nuclear Weapons Option, 1949–1999'. *The Nonproliferation Review* 7, no. 1 (2000): 1–23.

Information Office of the State Council of the People's Republic of China. *China's Non-Proliferation Policy and Measures*. Beijing: New Star Publishers, 2003.

———. *China's Endeavors for Arms Control, Disarmament and Non-Proliferation.* March 2005. Available at http://english.xinhua.cn.

International Atomic Energy Agency (IAEA). 'Power Reactor Information System'. Available at http://www.iaea.org/PRIS/WorldStatistics/OperationalReactorsByCountry.aspx.

———. *Convention on Nuclear Safety*, INFCIRC/440. Vienna: International Atomic Energy Agency, 1994. Available at http://www.iaea.org/Publications/Documents/Infcircs/Others/inf449.shtml.

———. *International Nuclear Safeguards 1994: Vision for the Future*, IAEA-SM-333/107. Vienna: International Atomic Energy Agency, 1995.

International Atomic Energy Agency (IAEA), Nuclear Knowledge Management Section. *The Impact of Knowledge Management Practices on NPP Organizational Performance – Results of a Global Survey*. IAEA-TECDOC-1711. Vienna: International Atomic Energy Agency, 2013.

International Commission on Nuclear Non-proliferation and Disarmament (ICNND). 'Joint Statement by Gareth Evans and Yoriko Kawaguchi'. New York, 25 September 2008. Available at http://www.mofa.go.jp/ICSFiles/afieldfile/2008/09/26/h2009_nichi_2.pdf.

———. *Eliminating Nuclear Threats: A Practical Agenda for Global Policymakers*. Canberra: Paragon, 2009.

International Court of Justice. *International Court of Justice Yearbook: 1993–1994*, no. 48. The Hague: International Court of Justice, 1994.

———. *International Court of Justice Yearbook: 1994–1995*, no. 49. The Hague: International Court of Justice, 1995.

Jacobsen, P. and M. Binyon. 'France Calls off Nuclear Testing'. *The Times*. 9 April 1992.

Johnson, Rebecca. 'Test Ban Treaty: Rearranging Deck Chairs on the Titanic'. *Bulletin of the Atomic Scientists* 51, no. 5 (September/October 1995): 11–12.

Johnston, Alastair Iain. 'Learning versus Adaptation: Explaining Change in Chinese Arms Control Policy in the 1980s and 1990s'. *The China Journal* 35 (January 1996): 27–61.

Joint Committee on Foreign Affairs and Defence. *Threats to Australia: Their Nature and Probability*. Canberra: Australian Government Publishing Service, 1981.

Jones, Bruce. 'ANZUS: Is it Worth the Paper It's Written On?' *The Sydney Morning Herald*. 30 April 1989.

Jones, David M. and Andrea Benvenuti. 'Tradition, Myth and the Dilemma of Australian Foreign Policy'. *Australian Journal of International Affairs* 60, no. 1 (2006): 103–24.

Jones, David M. and Mike L. Smith. 'Misreading Menzies and Whitlam: Reassessing the Ideological Construction of Australian Foreign Policy'. *The Round Table: The Commonwealth Journal of International Affairs* 89, no. 355 (2000): 387–406.

Jordan, Matthew. 'Arms Control and Disarmament'. In James Cotton and David Lee, eds, *Australia and the United Nations*. Canberra: Commonwealth of Australia, 2012.

Joseph, Jofi. 'The Exercise of National Sovereignty: The Bush Administration's Approach to Combating Weapons of Mass Destruction Proliferation'. *The Nonproliferation Review* 12, no. 2 (2005): 373–87.

Kaiser, Karl. 'The Great Nuclear Debate: German-American Disagreements'. *Foreign Policy* 30 (Spring 1978): 83–110.

Kakodkar, Anil. 'Managing Nuclear Knowledge: A Governmental Perspective'. *International Journal of Nuclear Knowledge Management* 1, no. 1/2 (2004): 33–8.

Kapur, Ashok. 'The Future of the NPT: A View from the Indian Subcontinent'. in D. Dewitt, ed., *Nuclear Non-Proliferation and Global Security*. New York: St. Martin's Press, 1987.

Karam, P. Andrew. 'How do Fast Breeder Reactors Differ from Regular Nuclear Power Plants?' *Scientific American*, 17 July 2006. Available at http://www.scientificamerican.com/article.cfm?id=how-do-fast-breeder-react.

Kaucher, L. 'Nuclear Debate Needed'. *The Sydney Morning Herald*. 15 September 1995.

Kay, David. 'Denial and Deception Practices of WMD Proliferators: Iraq and Beyond'. *Washington Quarterly* 18, no. 1 (Winter 1995): 85–105.

Keating, Paul. 'Australian Initiative for a Nuclear Weapons-Free World'. Statement by the Prime Minister, the Hon. P.J. Keating, MP, 26 November 1995. Available at http://pmtranscripts.dpmc.gov.au/transcripts/00009857.pdf.

Kelly, Paul. 'The Carter Connection'. *The National Times*. 30 May–4 June 1977.

Keohane, Robert O. 'Lilliputians' Dilemmas: Small States in International Politics'. *International Organization* 23, no. 2 (1969): 291–310.

Klare, Michael T. *Rogue States and Nuclear Outlaws: America's Search for a New Foreign Policy*. New York: Macmillan, 1996.

Koch, Susan. 'Proliferation Security Initiative: Origins and Evolution'. *Center for the Study of Weapons of Mass Destruction Occasional Paper No. 9*. Washington, DC: National Defense University Press, 2012. Available at http://www.ndu.edu/press/lib/pdf/CSWMD-OccasionalPapers/CSWMD_OccationalPaper-9.pdf.

Krause, Joachim. 'Enlightenment and Nuclear Order'. *International Affairs* 83, no. 3 (2007): 483–99.

Krepon, Michael. 'The Stability-Instability Paradox, Misperception, and Escalation-Control in South Asia'. In Raaq Dossani and Henry S. Rowen, eds, *Prospects for Peace in South Asia*. Stanford, CA: Stanford University Press, 2005.

Kristensen, Hans M. and Joshua Handler. 'The USA and Counter-Proliferation: A New and Dubious Role for US Nuclear Weapons'. *Security Dialogue* 27, no. 4 (1996): 387–99.

Kroenig, Matthew. 'Importing the Bomb: Sensitive Nuclear Assistance and Nuclear Proliferation'. *Journal of Conflict Resolution* 53, no. 2 (2009): 161–80.

Kubbig, Bernd W. 'America: Escaping the Legacy of the ABM Treaty'. *Contemporary Security Policy* 26, no. 3 (2005): 410–12.

Kurihara, H. 'Role of Support Programmes in Safeguards'. *IAEA Bulletin* 25, no. 1 (1988): 13–17.

Lantis, Jeffrey. 'Elections and Enduring Realities: Australia's Nuclear Debate'. *Arms Control Today* (April 2008). Available at https://www.armscontrol.org/act/2008_04/Lantis.

Laughter, M.D. *Profile of World Uranium Enrichment Programs, 2009*, ORNL/TM-2009/110. Oak Ridge, TN: Oak Ridge National Laboratory, 2009.

Lawrence, Robert M. and Joel Larus, eds. *Nuclear Proliferation Phase II*. Lawrence: University Press of Kansas, 1974.

Lax, Mark. *From Controversy to Cutting Edge: A History of the F-111 in Australian Service*. Canberra: Air Power Development Centre, 2010.

Leah, Christine. 'US Extended Deterrence and Nuclear Order: An Australian Perspective'. *Asian Security* 8, no. 2 (2012): 93–114.

——— and Rod Lyon. 'Three Visions of the Bomb: Australian Thinking about Nuclear Weapons and Strategy'. *Australian Journal of International Affairs* 64, no. 4 (2010): 449–74.

Levy, Adrian and Catherine Scott-Clark. *Deception: Pakistan, the United States and the Global Nuclear Conspiracy*. London: Atlantic Books, 2007.

Lewis, George, Lisbeth Gronlund and David Wright. 'National Missile Defense: An Indefensible System'. *Foreign Policy* (Winter 1999/2000): 120–31.

Lim, Robyn. 'Australian Security Thinking after the Cold War'. *Orbis: A Journal of World Affairs* 42, no. 1 (Winter 1998): 91–103.

Lindemeyer, Jeff. 'The Global Nuclear Energy Partnership: Proliferation Concerns and Implications'. *The Nonproliferation Review* 16, no. 1 (2009): 79–93.

Lockwood, D. 'US Begins Testing Moratorium, Ban on US Tests Set for 1996'. *Arms Control Today* 22, no. 8 (October 1992).

———. 'CD Negotiators Offer Proposals for Comprehensive Test Ban Treaty'. *Arms Control Today* 24, no. 3 (1994).

Lowy Institute for International Policy, Nuclear Policy Centre. Available at http://m.lowyinstitute.org/programs-and-projects/projects/nuclear-policy-centre.

Lyon, Rod. 'Australia: Back to the Future?' In Muthiah Alagappa, ed., *Security in 21st Century Asia*. Stanford, CA: Stanford University Press, 2008.

MacArthur, K. 'Australian Submission to the International Court of Justice on the Legality of the Use of Nuclear Weapons in Armed Conflict'. *Peace and Disarmament News* (August 1994).

MacLeod, Roy. 'The Atom Comes to Australia: Reflections on the Australian Nuclear Program, 1953 to 1993'. *History and Technology* 11, no. 2 (1994): 304–9.

MacMillan, Richard. *Aegis TMD: Implications for Australia*. Canberra: Australian Defence College, Monograph Series, no. 1, 2003.

Mahnken, Thomas G. 'A Critical Appraisal of the Defense Counterproliferation Initiative'. *National Security Studies Quarterly* 5, no. 3 (Summer 1999): 91–102.

Malik, Mohan. 'Nuclear Proliferation in Asia: The China Factor'. *Australian Journal of International Affairs* 53, no. 1 (1999): 31–41.

Martin, B. *Nuclear Knights*. Canberra: Rupert Public Interest Movement, 1980.

Martinez, J. Michael. 'The Carter Administration and the Evolution of American Nuclear Nonproliferation Policy, 1977–1981'. *Journal of Policy History* 14, no. 3 (2002): 261–92.

Massachusetts Institute of Technology. *The Future of Nuclear Power: An Interdisciplinary MIT Study*. Cambridge, MA: Massachusetts Institute of Technology, 2003. Available at http://web.mit.edu/nuclearpower/.

Matthews, Trevor and John Ravenhill. 'Bipartisanship in the Australian Foreign Policy Elite'. *Australian Outlook* 42, no. 1 (1988): 9–20.

McClelland, Robert. 'A New Agenda for Australia in Nuclear Non-Proliferation and Disarmament'. Address to UN Association of Australia and Medical Association for the Prevention of War. Canberra, 14 August 2007. Available at http://nautilus.org/apsnet/a-new-agenda-for-australia-in-nuclear-non-proliferation-and-disarmament/.

McDowall, Rory C. *Howard's Long March: The Strategic Depiction of China in Howard Government Policy, 1996–2006*. Canberra: ANU E-Press, 2009.

McKinnon, K.R., A. Henderson-Sellers and T. Hundloe. *Future Reaction: Report of the Research Reactor Review*. Sydney: Australian Government Publishing Service, 1993.

McLachlan, Ian. 'Australia and United States Cooperate on Missile Detection'. Media release by the Minister for Defence, 8 August 1997.

McLean, David. 'Australia in the Cold War: A Historiographical Review'. *The International History Review* 23, no. 2 (2001): 299–321.

Mian, Zia, A.H. Nayyar, R. Rajaraman and M.V. Ramana. 'Fissile Materials in South Asia: The Implications of the US-India Nuclear Deal'. *International Panel on Fissile Materials*, Research Report no. 1. September 2006. Available at http://www.fissilematerials.org/ipfm/site_down/rr01.pdf.

Middleton, Karen. 'Test Anger Misjudged, Says Evans'. *The Age*. 28 June 1995.

Millar, Thomas B. *Australia's Defence*, 2nd edition. Carlton: Melbourne University Press, 1969.

———. 'From Whitlam to Fraser'. *Foreign Affairs* 55, no. 4 (1977): 854–72.

Miller, Linda B. 'The Clinton Years: Reinventing US Foreign Policy?' *International Affairs* 70, no. 4 (1994): 621–34.

Milne, Christine. 'Election Silence on Uranium Exports Me-Too-Ism'. Australian Greens Media Release, 21 November 2007. Available at http://greens.org.au.

Missile Defense Agency. 'Project DUNDEE Fact Sheet'. Available at http://www.acq.osd.mil/bmdo/bmdolink/html/dndefs.html.

Monten, Jonathan and Mark Provost. 'Theater Missile Defense and Japanese Nuclear Weapons'. *Asian Security* 1, no. 3 (2005): 285–303.

Mulhall, Daniel. 'Australia and Disarmament Diplomacy, 1983–1985: Rhetoric or Achievement?' *Australian Outlook* 40, no. 1 (1986): 32–8.

Müller, Harald and Mitchell Reiss. 'Counter-proliferation: Putting New Wine in Old Bottles'. *The Washington Quarterly* 18, no. 2 (Spring 1995): 143–54.

Müller, Harald, David Fischer and Wolfgang Kötter, eds. *Nuclear Non-Proliferation and Global Order.* Oxford: Oxford University Press, 1994.

Murphy, Katherine. 'Howard's U-Turn on India'. *The Age.* 30 March 2007.

National Institute for Public Policy. *The Proliferation Security Initiative: A Model for Future International Collaboration.* Fairfax, VA: National Institute Press, 2009.

Natural Resources Defense Council. 'Nuclear Weapons Policy'. Available at http://www.nrdc.org/nuclear/nuguide/nwipoli.asp.

Newman, Andrew and Brad Williams. 'The Proliferation Security Initiative'. *The Nonproliferation Review* 12, no. 2 (2005): 303–22.

Nicholson, Brendan. 'Slaking a Giant's Thirst'. *The Australian.* 16 November 2011.

Nilsson, Måns. 'The Role of Assessments and Institutions for Policy Learning: A Study on Swedish Climate and Nuclear Policy Formation'. *Policy Sciences* 38, no. 4 (2005): 225–49.

Norman, J. 'New Risks from Australia-Russia Uranium Deal'. *Online Opinion.* 19 November 2010. Available at http://www.onlineopinion.com.au/view.asp?article=11251.

North Atlantic Council. *The Alliance's New Strategic Concept.* Brussels: NATO, 1991.

———. *The Alliance's Strategic Concept.* Brussels: NATO, 1999.

Nuclear Energy Agency. *Trends in Nuclear Research Institutes.* Paris: Organization for Economic Cooperation and Development, 1996.

———. *Nuclear Legislation in OECD and NEA Countries: Australia.* Paris: Organization for Economic Cooperation and Development, 2008.

———. *The Path to a Reliable Supply of Medical Radioisotopes: Nuclear Energy in Perspective.* Paris: Organization for Economic Cooperation and Development, 2011.

———. *GIF Symposium Proceedings: 2012 Annual Report.* Paris: Organization for Economic Cooperation and Development, 2012.

———. *Nuclear Education and Training: From Concern to Capability.* Paris: Organization for Economic Cooperation and Development, 2012.

———. *Nuclear Energy Today.* Paris: Organization for Economic Cooperation and Development, 2012.

Nuclear Industry Association. 'Nuclear Policy in the UK – Where are we now?' Available at http://www.niauk.org/nuclear-policy-in-the-uk.

Nye, Joseph S. 'Nonproliferation: A Long-Term Strategy'. *Foreign Affairs* 56, no. 3 (1978): 604–6.

———. 'NPT: The Logic of Inequality'. *Foreign Policy* 59 (Summer 1985): 123–31.

Oakes, Dan. 'Emirates Gets Nod to Buy Uranium'. *The Age.* 10 March 2011.

O'Brien, Jason. 'Canberra Yellowcake: The Politics of Uranium and How Aboriginal Land Rights Failed the Mirrar People'. *Journal of Northern Territory History* 14 (2003): 79–91.

Oda, Masakatsu. 'Japan Lobbied for more Nuclear Weapons'. *Kyodo News*. 23 November 2009.

O'Faircheallaigh, Ciaran. 'Uranium in Australia, 1970–1987: An Overview'. In Ciaran O'Faircheallaigh, Deborah Wade-Marshall and Ann Webb, eds, *Uranium in Australia, 1970–1987: An Annotated Bibliography*. Canberra: Australian National University, North Australia Research Unit, 1989.

Office of International Security Affairs. *United States Security Strategy for the East Asia – Pacific Region*. Washington, DC: Department of Defense, 1995.

O'Hanlon, Michael. 'Star Wars Strikes Back'. *Foreign Affairs* 78, no. 6 (1999): 68–82.

———. *Defense Policy Choices for the Bush Administration*, 2nd edition. Washington, DC: Brookings Institution Press, 2002.

O'Neil, Andrew. 'The Politics of Denuclearisation: Assessing the Canberra Commission Report'. *Current Affairs Bulletin* 73, no. 2 (August/September 1996): 28–30.

———. 'Constraints and Continuity: The Evolution of Australia's Non-proliferation Policy, 1975–96'. Unpublished Doctor of Philosophy Thesis, Flinders University, 1999.

———. 'Australian Participation in National Missile Defence: Looking Beyond the Horizon'. *AQ: Australian Quarterly* 73, no. 3 (2001): 19–23.

———. 'Australia and the South Pacific Nuclear Free Zone Treaty: A Reinterpretation'. *Australian Journal of Political Science* 39, no. 3 (2004): 567–83.

———. 'Shifting Policy in a Nuclear World: Australia's Non-Proliferation Strategy since 9/11'. In Carl Ungerer, ed., *Australian Foreign Policy in the Age of Terror*. Sydney: UNSW Press, 2008.

———. 'Nuclear Weapons and Non-proliferation: Is Restraint Sustainable?' *Security Challenges* 5, no. 4 (2009): 39–55.

———. 'Regional, Alliance and Global Priorities of the Rudd-Gillard Governments'. In James Cotton and John Ravenhill, eds, *Middle Power Dreaming: Australia in World Affairs, 2006–2010*. Melbourne: Oxford University Press, 2011. 273–89.

———. *Asia, the US and Extended Nuclear Deterrence: Atomic Umbrellas in the Twenty-First Century*. London and New York: Routledge, 2013.

———. 'Defence White Paper Pulls Its Punches on China'. *Lowy Interpreter*. 6 May 2013. Available at http://www.lowyinterpreter.org/post/2013/05/06/Defence-White-Paper-pulls-its-punches-on-China.aspx?COLLCC=3084967594&.

O'Neill, Kate. 'Radioactive "Trade": Globalizing the Nuclear Fuel Cycle'. *SAIS Review*, no. 1 (Winter-Spring 2002): 157–68.

O'Neill, Robert. *Internal Problems Associated with any Acquisition of Nuclear Weapons by Australia*. Canberra: Strategic and Defence Studies Centre, Australian National University, 1974.

———, ed. *The Strategic Nuclear Balance: an Australian Perspective*. Canberra: Strategic and Defence Studies Centre, Australian National University, 1974.

Packham, B. 'Nuclear Deal with Russia put on Ice'. *The Herald Sun*. 19 September 2008.

Parliamentary Joint Standing Committee on Treaties. *Inquiry into Nuclear Non-proliferation and Disarmament*, Report no. 106. Canberra, Australian Government Publishing Service, September 2009.

Paul, T.V. 'Systemic Conditions and Security Cooperation: Explaining the Persistence of the Nuclear Non-proliferation Regime'. *Cambridge Review of International Affairs* 16, no. 1 (2003): 135–54.

Payne, Keith B. and Colin S. Gray. 'Nuclear Policy and the Defensive Transition'. *Foreign Affairs* 62, no. 4 (1984): 820–42.

Pellaud, Bruno. 'Safeguards: The Evolving Picture'. *IAEA Bulletin* 38, no. 4 (1996): 2–6.

Percebois, Jacques. 'The Peaceful Uses of Nuclear Energy: Technologies of the Front and Back-Ends of the Fuel Cycle'. *Energy Policy* 31 (2003): 101–8.

Perrett, J. 'Nuclear Pact Wins Soviet Praise'. *The Weekend Australian*. 28–29 September 1985.

Perry, William J. and Brent Scowcroft (chairs). 'U.S. Nuclear Weapons Policy'. *Independent Task Force* Report no. 62. Washington, DC: Council on Foreign Relations Press, 2009.

Pidd, Helen and Suzanne Goldenberg. 'Germany suspends power station extension plans as nuclear jitters spread'. *The Guardian*. 15 March 2011. Available at http://www.theguardian.com/environment/2011/mar/14/germany-japan-nuclear-industry?guni=Article:in%20body%20link.

Podvig, Pavel. *Russian Strategic Nuclear Forces*. Cambridge, MA: MIT Press, 2004.

Pollack, Jonathan. *No Exit: North Korea, Nuclear Weapons and International Security*. London: Routledge and IISS, 2011.

Poulose, T.T. 'Nuclear Proliferation: A Third World Perspective'. *The Round Table* 69, no. 274 (1979): 154–68.

Power, Paul F. 'The Indo-American Nuclear Controversy'. *Asian Survey* 19, no. 6 (1979): 574–96.

Reagan, Ronald. 'Statement on United States Nuclear Nonproliferation Policy'. 16 July 1981. *Public Papers of President Ronald Reagan*. Available at http://www.reagan.utexas.edu/archives/speeches/1981/71681a.htm.

Reiss, Mitchell. *Bridled Ambition: Why Countries Constrain Their Nuclear Capabilities*. Washington, DC: Woodrow Wilson Center Press, 1995.

Reith, Peter. 'Address by the Hon. Peter Reith, MP, Minister for Defence to the ANZUS Conference'. 30 June 2001. Available at http://www.defence.gov.au/minister/8tpl.cfm?CurrentId=785.

Reus-Smit, C. 'Arms Control, Nuclear Strategy and Australian Foreign Policy: The Fraser Years'. *Interdisciplinary Peace Research* 1, no. 2 (October/November 1989): 57–73.

Reynolds, David. 'Empire, Region, World: The International Context of Australian Foreign Policy since 1939'. *Australian Journal of Politics and History* 51, no. 3 (2005): 346–58.

Reynolds, Wayne. *Australia's Bid for the Atomic Bomb*. Melbourne: Melbourne University Press, 2000.

———. 'Australia's Quest to Enrich Uranium and the Whitlam Government's Loans Affair'. *Australian Journal of Politics and History* 54, no. 4 (2008): 562–78.

———. 'The Yellow Cake Road: Malcolm Fraser, the Ranger Enquiry and Australia's Role in the US International Fuel Cycle Project'. *Australian Journal of Politics and History* 57, no. 4 (2011): 511–25.

——— and David Lee, eds. *Australia and the Nuclear Non-Proliferation Treaty, 1945–1974*. Canberra: Department of Foreign Affairs and Trade, 2013.

Richardson, J.L. 'Australia and the Non-Proliferation Treaty'. *Canberra Papers on Strategy and Defence*, no. 3. Canberra: Australian National University Press, 1968.

Richelson, Jeffrey T. *Spying on the Bomb: American Nuclear Intelligence from Nazi Germany to Iran and North Korea*. New York: W.W. Norton & Company, 2006.

Risse-Kappen, Thomas. 'Did "Peace through Strength" End the Cold War? Lessons from INF'. *International Security* 16, no. 1 (Summer 1991): 162–88.

Rockwood, Laura. 'Safeguards and Nonproliferation: The First Half-Century from a Legal Perspective'. *Journal of Nuclear Materials Management* 35, no. 4 (2007): 79–93.

Rose, Frank A. 'The Future of U.S. Nuclear Policy'. Washington, DC: The Pryzbyla Center, Catholic University of America, 9 April 2013. Available at http://www.state.gov/t/avc/rls/2013/207360.htm.

Ross, R.S. 'The 1995–96 Taiwan Strait Confrontation: Coercion, Credibility, and the Use of Force'. *International Security* 25, no. 2 (2000): 87–123.

Rostow, Nicholas. 'The World Health Organization, the International Court of Justice, and Nuclear Weapons'. *Yale Journal of International Law* 20, no. 1 (Winter 1995): 151–85.

Rublee, Maria Rost. *Nonproliferation Norms: Why States Choose Nuclear Restraint*. Athens and London: University of Georgia Press, 2009.

Sagan, Scott D. 'Why Do States Build Nuclear Weapons? Three Models in Search of a Bomb'. *International Security* 21, no. 3 (Winter 1996/97): 63–4.

———. 'Nuclear Latency and Nuclear Proliferation'. In William Potter, ed., *Forecasting Nuclear Proliferation in the 21st Century*. Stanford, CA: Stanford University Press, 2010.

Schapiro, M. 'Mutiny on the Nuclear Bounty'. *The Nation* (27 December 1993): 798–800.

Scheinman, Lawrence. *The International Atomic Energy Agency and World Nuclear Order*. Washington, DC: Resources for the Future, 1987.

Schell, Jonathan. 'The Second Age of Nuclear Danger'. *Boston Review* (April/May 1998).

Schneider, Mycle and Antony Froggatt. *The World Nuclear Industry Status Report 2012*. Paris: Mycle Schneider Consulting, 2012.

Schofield, Clive, Ben M. Tsamenyi and Mary Anna Palma. 'Securing Maritime Australia: Developments in Maritime Surveillance and Security'. *Ocean Developments and International Law* 39, no. 1 (2008): 94–112.

Schou, A. and A.O. Brundtland, eds. *Small States in International Relations*. Stockholm: Almqvist and Wiksell, 1971.

Schriefer, Dirk. 'New Safeguards Measures: Initial Implementation and Experience'. *IAEA Bulletin* 4 (1996): 7–10.

Schubert, Mischa. 'PMs Vision of Australia as Energy Superpower'. *The Age*. 18 July 2006. Available at http://www.theage.com.au/news/business/pms-vision-of-australia-as-energy-superpower/2006/07/17/1152988472470.html.

Schultz, George, William Perry, Henry Kissinger and Sam Nunn. 'A World Free of Nuclear Weapons'. *The Wall Street Journal*. 4 January 2007.

Seth, S. 'China and India's nuclear energy plans to drive global uranium demand'. *Mineweb.com*. 9 December 2010. Available at http://www.mineweb.com/mineweb/view/mineweb/en/page72103?oid=116592&sn=Detail.

Sharma, D. 'Uranium Trade and its Security Implications for India'. *South Asian Survey* 17, no. 1 (2010): 91–110.

Sheridan, G. 'Disarmament Talk Endangers US Alliance'. *The Australian*. 24 January 1996.

Silex. *Annual Report 2013*. Available at http://www.silex.com.au/downloads/asxannouncements/17-slx-annual-report-2013-fa-online.

Simons, Howard. 'World-Wide Capabilities for Production and Control of Nuclear Weapons'. *Daedalus* 88, no. 3 (1959): 385–409.

Simpson, Christopher. *National Security Directives of the Reagan and Bush Administrations: The Declassified History of US Political and Military Policy, 1981–1991*. Boulder, CO: Westview Press, 1995.

Slatyer, A.O. *Australia's Role in the Nuclear Fuel Cycle: A Report to the Prime Minister by the Australian Science and Technology Council (ASTEC)*. Canberra: Australian Government Publishing Service, 1984.

Smith, Gerard and George Rathjens. 'Reassessing Nuclear Nonproliferation Policy'. *Foreign Affairs* 59, no. 4 (1981): 880–82.

Smith, R.J. 'Administration at Odds over Soviet Cheating'. *Science* 228, no. 4700 (10 May 1985): 695–6.

Smith, Stephen. 'Building Momentum: Australia, Nuclear Non-Proliferation and Disarmament'. Sir Arthur Tange Lecture in Australian Diplomacy. 12 August 2009. Available at http://www.foreignminister.gov.au/speeches/2009/090812_tange.html.

———. 'Australian Government's Response to the International Commission on Nuclear Non-proliferation and Disarmament Report'. 3 May 2010. Available at http://www.foreignminister.gov.au/releases/2010/fa-s100503a.html.

Sorokowsi, D.L., K. Veevers and C.S. Johnson. 'Co-operative activities of the Australian Safeguards Office and Sandia National Laboratories in Advanced Containment and Surveillance Technologies (IAEA-SM-333/107)'. In *International Nuclear Safeguards 1994: Vision for the Future*. Vienna: IAEA, 1994. 631–40.

Squassoni, Sharon. *Nuclear Energy: Rebirth or Resuscitation?* Washington, DC: Carnegie Endowment for International Peace, 2009.

Steketee, M. 'Hawke Advisers Drew N-Arms Plan'. *The Weekend Australian*. 6–7 April 1996.

Stockholm International Peace Research Institute. 'The Implementation of Agreements Related to Disarmament'. In Stockholm International Peace Research Institute, ed., *World Armaments and Disarmament: SIPRI Yearbook 1976*. Stockholm. Cambridge, MA: MIT Press, 1976. 363–92.

———. *World Armaments and Disarmament: SIPRI Yearbook 1982*. London: Taylor and Francis, 1982.

Stoessinger, John. 'Atoms for Peace: The International Atomic Energy Agency'. In Commission to Study the Organization of Peace, *Organizing Peace in the Nuclear Age*. New York: New York University Press, 1959.

Stoiber, Carlton, Abdelmadjid Cherf, Wolfram Tonhauser and Maria de Lourdes Vez Carmona. *Handbook on Nuclear Law: Implementing Legislation*. Vienna: IAEA, 2010.

Sun, L. 'South Pacific Nations Declare Nuclear-Free Zone'. *The Washington Post*. 8 August 1985.

Suter, Keith D. 'The 1975 Review Conference of the Nuclear Non-Proliferation Treaty'. *Australian Outlook* 30, no. 2 (August 1976): 322–40.

Takubo, Masa. 'The Role of Nuclear Weapons: Japan, the U.S., and "Sole-Purpose"'. *Arms Control Today* (November 2009). Available at https://www.armscontrol.org/act/2009_11/Takubo#7.

Tanter, Richard. 'Just in Case: Extended Nuclear Deterrence in the Defence of Australia'. *Pacific Focus* 26, no. 1 (2011): 113–36.

———. 'American Bases in Australia Revisited'. In Brendan Taylor, Nicholas Farrelly and Sheryn Lee, eds, *Insurgent Intellectual: Essays in Honour of Professor Desmond Ball*. Singapore: ISEAS Publishing, 2012. 191–211.

Taylor, Brendan, Nicholas Farrelly and Sheryn Lee, eds. *Insurgent Intellectual: Essays in Honour of Professor Desmond Ball*. Singapore: ISEAS Publishing, 2012.

Taylor, M. 'The Day We Tried to Get the Bomb'. *The Australian*. 1 January 1988.

Tellis, Ashley. 'The Evolution of US-Indian Ties: Missile Defense in an Emerging Strategic Relationship'. *International Security* 30, no. 4 (2006): 113–51.

Thakur, Ramesh. 'The Differing No-Nuke Stance between Australia and New Zealand'. *Worldviews*. 29 November 2013. Available at http://passblue.com/2013/11/29/the-differing-no-nuke-stance-between-australia-and-new-zealand/.

———. 'Follow the Yellowcake Road: Balancing Australia's National Interests against International anti-Nuclear Interests'. *International Affairs* 89, no. 4 (2013): 943–61.

Thompson, Alan. *Defence Down Under: Evolution and Revolution 1971–88*, Working Paper 40. London: Sir Robert Menzies Centre for Australian Studies, University of London, 1988.

Tisco, Antonio. 'The Carter Administration and its Non-Proliferation Policies: the Road to INFCE'. *Humana Mente: Journal of Philosophical Studies* 16 (2011): 60–64.

Toohey, Brian. *Book of Leaks*. Sydney: Angus & Robertson, 1987.

Tow, William. 'Sino-American Relations and the "Australian Factor": Inflated Expectations or Discriminate Engagement?' *Australian Journal of International Affairs* 59, no. 4 (December 2005): 451–67.

——— and Leisa Hay. 'Australia, the United States and a "China Growing Strong": Managing Conflict Avoidance'. *Australian Journal of International Affairs* 55, no. 1 (2001): 37–54.

Towle, P. 'Australian Policy in the Committee on Disarmament'. *Strategic and Defence Studies Centre Working Paper*, no. 18. Canberra: Strategic and Defence Studies Centre, Australian National University, 1980.

Trinca, Helen. 'US and Britain Argue French Will Sign Ban Treaty: Envoy'. *The Australian*. 21 July 1995.

Trood, Russell. 'Australian Uranium Exports: Nuclear Issues and the Policy Process'. In Robert Boardman and James F. Keeley, eds, *Nuclear Exports and World Politics: Policy and Regime*. London: Palgrave Macmillan, 1983.

———. 'Prime Ministers and Foreign Policy'. In Patrick Weller, ed., *Menzies to Keating: The Development of the Australian Prime Ministership*. Carlton: Melbourne University Press, 1992.

Tucker, Jonathan B. 'Monitoring and verification in a non-cooperative environment: Lessons from the U.N. experience in Iraq'. *The Nonproliferation Review* 3, no. 3 (Spring-Summer 1996): 1–14.

Tzeng, Peter. 'Nuclear Leverage'. *The Nonproliferation Review* 20, no. 3 (2013): 473–92.

Ungerer, Carl. 'The Middle Power Concept in Australian Foreign Policy'. *Australian Journal of Politics and History* 53, no. 4 (2007): 538–51.

———. 'Australia's Place in the International System: Middle Power, Pivotal Power or Dependent Power?' In Carl Ungerer, ed., *Australian Foreign Policy in the Age of Terror*. Sydney: UNSW Press, 2008.

——— and Marianne Hanson, eds. *The Politics of Nuclear Non-Proliferation*. St. Leonards, NSW: Allen and Unwin, 2001.

United Nations. *Report of the Committee on Disarmament, 1980: General Assembly Official Records, Thirty-Fifth Session, Supplement*, no. 27 (A/35/27). New York: United Nations, 1980.

———. *Yearbook of the United Nations: 1981* 35. New York: United Nations Department of Public Information, 1985.

———. *Status of Multilateral Arms Regulation and Disarmament Agreements*, 3rd edition. New York: United Nations Department for Disarmament Affairs, 1988.

———. *Yearbook of the United Nations 1994* 48. New York: United Nations Department of Public Information, 1995.

United Nations Office for Disarmament Affairs. 'Australia Puts Forward Draft Treaty Elements'. *Disarmament: Newsletter of the UN Centre for Disarmament Affairs* 12, no. 2 (1994): 3.

United Nations Office of the Spokesperson for the Secretary-General. 'Kofi Annan Says Failure to Include Non-proliferation in Summit Document is a "Real Disgrace"'. 13 September 2005. Available at http://www.un.org/sg/spokesperson/highlights/?HighD=9/13/2005&d_month=9&d_year=2005.

United States Congress Commission to Assess the Ballistic Missile Threat to the United States. *Executive Summary of the Commission to Assess the Ballistic Missile Threat to the United States.* 15 July 1998. Available at http://fas.org/irp/threat/bm-threat.htm.

United States Congress Office of Technology Assessment. *Technologies Underlying Weapons of Mass Destruction*, OTA-BP-ISC-115. Washington, DC: United States Government Printing Office, 1993.

United States Counterproliferation Program Review Committee. *Report on Activities and Programs for Countering Proliferation and NBC Terrorism, Executive Summary.* Washington, DC: Government Printing Office, October 2001. Available at http://www.nti.org/media/pdfs/40_5.pdf?=1318280983.

United States Department of Defense. *The 41st US-ROK Security Consultative Meeting Joint Communique.* Seoul, South Korea, 2009. Available at http://www.defense.gov/releases/release.aspx?releaseid=13072.

———. *Nuclear Posture Review Report.* Washington, DC: United States Government Printing Office, 2010.

United States Department of State. *Treaty on the Non-Proliferation of Nuclear Weapons: U.S. Delegation to the 2010 Nuclear Nonproliferation Treaty Review Conference.* Available at http://www.state.gov/documents/organization/141503.pdf.

United States Secretary of Defense. *Annual Report to the President and the Congress.* Washington, DC: Government Printing Office, 1994.

Utgoff, Victor A., ed. *The Coming Crisis: Nuclear Proliferation, US Interests and World Order.* Cambridge, MA: MIT Press, 2000.

———. 'The Coming Crisis: Nuclear Proliferation, US Interests, and World Order – A Combined Perspective'. In Victor A. Utgoff, ed., *The Coming Crisis: Nuclear Proliferation, US Interests and World Order.* Cambridge, MA: MIT Press, 2000.

Valencia, Mark J. 'The Proliferation Security Initiative: Making Waves in Asia'. *The Adelphi Papers* 45, no. 376 (2005): 25–38.

von Hippel, Frank. 'Plutonium and Reprocessing of Spent Nuclear Fuel'. *Science* 5539, no. 28 (2001): 2397–8.

Walker, J. Samuel. 'Nuclear Power and Nonproliferation: The Controversy over Nuclear Exports, 1974–1980'. *Diplomatic History* 25, no. 2 (2001): 215–49.

——— and Thomas R. Wellock. *A Short History of Nuclear Regulation, 1946–2009*. Washington, DC: US Nuclear Regulatory Commission, 2010.

Walker, William. 'International Nuclear Relations after the Indian and Pakistani Test Explosions'. *International Affairs* 74, no. 3 (1998): 505–28.

———. 'Nuclear Order and Disorder'. *International Affairs* 76, no. 4 (2000): 703–24.

———. 'Nuclear Enlightenment and Counter-enlightenment'. *International Affairs* 83, no. 3 (2007): 431–53.

———. *A Perpetual Menace: Nuclear Weapons and International Order*. London: Routledge, 2012.

Walsh, Jim. 'Surprise Down Under: The Secret History of Australia's Nuclear Ambitions'. *The Nonproliferation Review* 5, no. 1 (1997): 1–20.

Walt, Stephen. 'Containing Rogues and Renegades: Coalition Strategies and Counter-Proliferation'. In Victor A. Utgoff, ed., *The Coming Crisis: Nuclear Proliferation, US Interests and World Order*. Cambridge, MA: MIT Press, 2000.

Warnecke, Steven J. 'Non-proliferation and the INFCE: An Interim Assessment'. *Survival* 21, no. 3 (1979): 116–24.

Wesley, Michael. 'Setting and Securing Australia's National Interests: The National Interest as Security'. In Ian Marsh, ed., *Australia's Choices: Options for a Prosperous and Fair Society*. Sydney: UNSW Press, 2003. 178–92.

———. *The Howard Paradox: Australian Diplomacy in Asia from 1996–2006*. Sydney: ABC Books, 2007.

——— and Tony Warren. 'Wild Colonial Ploys? Currents of Thought in Australian Foreign Policy'. *Australian Journal of Political Science* 35, no. 1 (2000): 9–26.

White, Andrew. 'European Perspectives on the Strategic Defense Initiative'. *Millennium-Journal of International Studies* 15, no. 2 (1986): 211–22.

White, Hugh. 'Power Shift: Rethinking Australia's Place in the Asian Century'. *Australian Journal of International Affairs* 65, no. 1 (2011): 81–93.

The White House Office of the Press Secretary. 'Remarks by the President to Students and Faculty at National Defense University'. Fort Lesley J. McNair, Washington, DC, 1 May 2001. Available at http://www.whitehouse.gov/news/releases/2001/05/20010501.html.

———. 'President, Prime Minister Singh Discuss Growing Strategic Partnership'. 2 March 2006. Available at http://georgewbushwhitehouse.archives.gov/news/releases/2006/03/images/20060302-9_p030206pm-0310-515h.html.

———. 'Remarks by President Obama'. Hradcany Square, Prague, Czech Republic, 5 April 2009. Available at http://www.whitehouse.gov/the_press_office/Remarks-By-President-Barack-Obama-In-Prague-As-Delivered.

Whitlam, Edward G. 'Australian Labor Party Policy Speech'. Blacktown Civic Centre, NSW. 13 November 1972. Available at http://electionspeeches.moadoph.gov.au/speeches/1972-gough-whitlam.

———. *The Whitlam Government: 1972–1975*. Ringwood: Penguin Books, 1985.

Wilkins, Thomas. 'Australia: A Traditional Middle Power Faces the Asian Century'. In Bruce Gilley and Andrew O'Neil, eds, *Middle Powers and the Rise of China*. Washington, DC: Georgetown University Press, 2014.

Wilrich, Mason. 'A Workable International Nuclear Energy Regime'. *Washington Quarterly* 2, no. 2 (1979): 13–30.

Winner, Andrew C. 'The Proliferation Security Initiative: The New Face of Interdiction'. *The Washington Quarterly* 28, no. 2 (2005): 129–43.

Wittner, Lawrence. *Resisting the Bomb: A History of the World Nuclear Disarmament Movement, 1954–1970*. Stanford, CA: Stanford University Press, 1997.

World Nuclear Association. 'Nuclear Power in the World Today'. Available at http://www.world-nuclear.org/info/Current-and-Future-Generation/Nuclear-Power-in-the-World-Today/.

———. 'Research Reactors'. Available at http://www.world-nuclear.org/info/Non-Power-Nuclear-Applications/Radioisotopes/Research-Reactors/#.UcPpXvlkPng.

World Nuclear News. 'Nuclear Policies'. Available at http://www.world-nuclear-news.org/sectionhub.aspx?fid=800.

X. 'Australian Doubts on the Treaty'. *Quadrant* 12, no. 3 (May/June 1968): 30–34.

Yoshihara, Toshi and James Holmes. 'Thinking about the Unthinkable: Tokyo's Nuclear Option'. *Naval War College Review* 62, no. 3 (2009): 59–78.

Index

AAEC. *See* Australian Atomic Energy Commission (AAEC)
Abbott government, 191
ACEL, 21
Additional Protocol, 112, 125, 138, 147–8, 159, 188
Age, The, 183
Alamogordo, 15
Algeria, 55
ALP. *See* Australian Labor Party (ALP)
americium-241, 19
ANSTO. *See* Australian Nuclear Science and Technology Organisation (ANSTO)
Anthony, Doug, 84, 85
Anti-Ballistic Missile Treaty, 56, 152
ANZUS. *See* Australia – New Zealand – United States Treaty (ANZUS)
A.Q. Khan network, 73n79, 141, 154
Areva, 21
Argentina, 25, 55
arms control, 25, 26, 99–103
ASADPO. *See Australian Strategic Analysis and Defence Policy Objectives* (ASADPO)
Asia-Pacific Safeguards Network, 148, 190
ASNO. *See* Australian Safeguards and Non-Proliferation Office (ASNO)
ASO. *See* Australian Safeguards Office (ASO)
ASTEC. *See* Australian Science and Technology Council (ASTEC)
Atomic Energy Act 1953, 39, 110, 111, 112
AUSMIN. *See* Australia-United States Ministerial (AUSMIN)
Australia
- Atomic Energy Policy Committee, 39
- Defence Committee, 34, 57, 61, 62, 97
- Department of Defence, 37, 95–6, 176
- Department of External Affairs, 34, 37
- Department of Foreign Affairs and Trade (DFAT), 95, 127, 136, 156, 162–3
- Department of National Development, 39, 111
- Department of Supply, 36, 39
- Department of Trade and Industry, 43
- Industrial Atomic Energy Policy Committee, 39
- Joint Intelligence Organization, 58
- research facilities of, 8, 39, 40, 163, 191, 197
- Safeguards Support Program, 8, 163, 197
- UK, relationship with, 4
- uranium resources of, 21, 31, 49–50, 78, 193
- US, relationship with, 4, 34–5, 143–4, 150, 152, 156, 191–2
- weapons development decision by, 31–8, 44, 95–9, 192, 196–7

'Australia and the World Situation' (Fraser), 58
Australia – New Zealand – United States Treaty (ANZUS), 91, 119, 144
Australia-United States Ministerial (AUSMIN), 120
Australian Atomic Energy Commission (AAEC), 8, 36, 39, 40, 50, 110, 163, 191
Australian Defence (1976 White Paper), 61, 96
Australian Labor Party (ALP), 4, 104, 107, 169, 178, 179, 184
Australian National University, 39, 96
Australian Nuclear Science and Technology Organisation (ANSTO), 8, 110–11, 127, 163, 191
Australian Safeguards and Non-Proliferation Office (ASNO), 127, 148
Australian Safeguards Assistance Program (ASAP), 126, 148
Australian Safeguards Office (ASO), 111, 112, 126, 127
Australian Science and Technology Council (ASTEC), 104, 105
Australian Strategic Analysis and Defence Policy Objectives (ASADPO), 59, 61
Australian Strategic Policy, 121, 124
Australia's Regional Security, 117, 121–2

Ball, Desmond, 60
Baxter, Sir Philip, 36, 44, 110
Beaufre, André, 18
Beazley, Kim, 94, 95, 99, 103, 109
Belgium, 41
beryllium-oxide moderated high-temperature gas-cooled reactor, 40
BHP, 84
Boiling Water Reactors (BWR), 20
Bolte, Sir Henry, 43
Booker, Malcolm, 34
Brazil, 25, 55, 71, 72
Bull, Hedley, 32
Bush, George W., 151–2, 154
Bush administration (US), 141, 151–2, 154, 157–8, 171
Butler, Richard, 91, 136
BWR, 20

Cameco, 21
Canada, 6, 8, 25, 71
  reactors in, 20, 41
  uranium resources of, 21
Canberra Commission on the Elimination of Nuclear Weapons (1995), 6, 128, 132, 135, 136–8, 171
Carnegie Endowment for International Peace, 12
Carr, Andrew, 170
Carr, Michael, 33
Carrick, John, 84
Carter, Jimmy, 64, 72, 73–4, 75, 77, 85
Carter administration (US), 68, 69–70, 73, 84, 194
Cawte, Alice, 44
CD. *See* Conference on Disarmament (CD)
CE-Hitachi, 21
Center for Strategic and International Studies, 11
Chemical Weapons Convention, 115
Chifley government, 38–9
China, 25, 56, 63, 117, 172, 174, 175
  export of uranium to, 158–60
  reactors in, 20, 21, 177
  strategic concerns, 17, 143
  testing by, 130, 144, 145
  treaty ratification by, 55, 115
  weapons in, 16, 17, 32, 35, 62, 128
Christopher, Warren, 84
Clinton, Hillary, 175
Clinton administration (US), 118, 122–3
cobalt-60, 19

Cold War, 16, 17, 18, 91, 93, 116–17
Comparison of the balance of forces between China and the U.S.S.R., 58
Comprehensive Nuclear Test Ban Treaty Act of 1998, 112
Comprehensive Test Ban Treaty (CTBT), 6, 24, 49, 64–5, 67–8, 100, 128–31, 146, 149, 150, 188
Conference on Disarmament (CD), 64, 68, 100, 130, 131, 146
Connor, Reginald 'Rex', 50
Convention for the Physical Protection of Nuclear Material, 112
Convention on Nuclear Safety (1994), 24
Converdyn, 21
Cooper, David, 123
Cotton, Robert, 43–4
Council on Foreign Relations, 12
counterproliferation, 122–3, 156
CSR, 84
CTBT. *See* Comprehensive Test Ban Treaty (CTBT)

Defence of Australia, The (1987 White Paper), 93, 99
Defence White Paper 2013, 176
Defending Australia: Defence White Paper 1994, 118, 119, 120, 121, 124, 144
Defending Australia in the Asia Pacific Century: Force 2030 (2009 White Paper), 174–5
'Des Ball: An Australian Original' (Dobbell), 61n24
deterrence, 1, 15, 16–17, 23, 26, 60, 93, 188
  extended, 6, 17, 34–5, 116–21, 174–6, 189, 192
  minimum, 17
Dibb Review (1986), 99
disarmament, 1, 26, 64, 131–8, 170–74, 192
Disarmament and Arms Control in the Nuclear Age (1986), 93
Dobbell, Graeme, 61n24
Downer, Alexander, 142–3, 145, 146, 147, 148, 149, 150, 152, 153–4, 156, 157, 160, 165

economic domain, 1. *See also* reactors, nuclear
  nuclear policy, 18–22, 28
  state involvement in, 25–6, 27
Einstein, Albert, 14–15
Eisenhower administration (US), 40, 41
Electricité de France, 20
Electrolytic Zinc Company, 51
Eliminating Nuclear Threats: A Practical Agenda for Global Policymakers, 173
energy, nuclear, 1, 12, 19, 27, 72–3. *See also* reactors, nuclear
  Australia's decision not to develop, 38–44
Enrichment Corporation, 21
Environmental Protection (Impact of Proposals) Act, 51, 53
epistemic communities, 22
EURATOM, 25, 55
Eurodif, 21
Evans, Chris, 178–9
Evans, Gareth, 93, 99, 106, 109, 110–11, 112, 115, 117, 121, 129, 132, 133, 134, 161, 171, 174, 180
Evatt, H.V., 39
exports of uranium, 6, 70, 74–80, 157–61, 163–7, 177–81, 189, 196. *See also* 'grand bargain'
  ASTEC review of, 104–7
  to China, 158–60
  to France, 104, 106
  to India, 157–8, 160, 170, 177–8, 181, 182, 183–5, 196
  to Russia, 180
extended nuclear deterrence, 6, 17, 34–5, 116–21, 174–6, 189, 192

Fairbairn, David, 37, 47
Fast Breeder Reactors, 20, 21–2, 72–3
Findlay, Trevor, 129
Finnemore, Martha, 7
Fissile Material Cut-off Treaty, 147
flexible response, 16, 35, 63
Ford, Gerald, 13–14
Ford administration (US), 69, 70, 71
Fox, Russell, 51
Fox Commission. *See* Ranger Uranium Environmental Inquiry
Framatome, 20
France, 18, 25, 41, 71, 72, 84, 107, 154, 172
  reactors in, 20, 21, 25
  testing by, 101, 129, 130
  treaty ratification by, 55, 115
  uranium exports to, 104, 106
  weapons in, 16, 62, 128
Fraser, Malcolm, 55, 57, 58, 60, 64, 73–4, 75–6, 77–8, 86, 87
Fraser government, 2, 9, 52, 55–89, 126, 141
  CTB promotion by, 63–9, 128
  global balance during, 56–61
  INFCE participation by, 80–83
  non-proliferation concerns during, 61–3
  nuclear proliferation in, 69–73
  security concerns during, 56–63
  uranium enrichment discussion in, 83–7, 110
  uranium exports during, 73–80
Frisch, Otto, 14
Fukushima disaster, 177
Furlonger, R.W., 49

gas-cooled reactors, 20
General Electric, 165
Generation IV, 25, 167
Germany, 15, 130, 154, 171
  reactors in, 20, 125
  treaty ratification by, 55
Gillard, Julia, 169, 170, 180, 183–4
Gillard government. *See* Rudd-Gillard-Rudd governments
Global Nuclear Fuel, 21
Gorbachev, Mikhail, 93
Gorton, John, 5, 37, 43, 47
Gorton government, 33, 35, 46, 47, 192
'grand bargain', 3, 7, 9, 104–7, 157, 181, 182, 193–5
Greenpeace, 165
growth of Soviet naval power in relation to that of the U.S., The, 58
Gulf War (1991), 116, 125
Gurry, Meg, 5

Hahn, Otto, 14
Hanson, Marianne, 6–7
Hasluck, Paul, 47
Hawke, Robert (Bob), 87, 91, 93, 94, 97, 105, 115
Hawke government, 4, 9, 91–113
  arms control during, 99–103
  domestic nuclear option during, 95–9
  non-proliferation efforts during, 109–12, 128
  nuclear arms control during, 190
  safeguards for non-NPT states during, 107–9, 126
  superpower arms race during, 92–5
  uranium exports during, 104–7
Hayden, Bill, 94, 97, 99, 103, 107, 109, 115, 134
hedging, 27
Heritage Foundation, 11
Hexapartite Safeguards Project, 8, 126, 197
HIFAR. *See* High Flux Australian Reactor (HIFAR)
Higgie, Dell, 174
High Flux Australian Reactor (HIFAR), 40, 161–3
Hiroshima, Japan, 15

Holst, J.J., 36
Howard, John, 141, 149, 153, 154, 157, 160, 164, 165
Howard government, 10, 44, 141–68, 169
  economic domain under, 161–7
  non-proliferation efforts during, 145–50
  strategic stability during, 142–5
  terrorism, response to, 151–6
  uranium exports during, 157–61, 177, 178
Hymans, Jacques, 5–6

IAEA. *See* International Atomic Energy Agency (IAEA)
ICBM. *See* Intercontinental Ballistic Missiles (ICBM)
ICJ. *See* International Court of Justice (ICJ) case
ICNND. *See* International Commission on Nuclear Non-proliferation and Disarmament (ICNND)
India, 55, 63, 117, 158
  nuclear capability of, 17, 141
  peaceful nuclear explosion (PNE), 17, 70, 72, 193–4
  reactors in, 20, 21, 177
  relationship with, 181–6
  strategic concerns, 17
  testing by, 149
  uranium export to, 157–8, 160, 170, 177–8, 181, 182, 183–5, 196
Indonesia, 32, 35, 55, 56, 62n29
INF. *See* Intermediate-Range Nuclear Forces (INF)
INFCE. *See* International Nuclear Fuel Cycle Evaluation (INFCE)
INFCIRC/153, 53, 111, 125
Intercontinental Ballistic Missiles (ICBM), 16, 56
Intermediate-Range Nuclear Forces (INF), 23
International Atomic Energy Agency (IAEA), 8, 23, 41, 116, 127, 149, 197
  Additional Protocol, 112, 125, 138, 147–8, 159, 188
  Asian Regional Cooperative Agreement (RCA) in Nuclear Science and Technology, 80
  Network of Analytical Laboratories, 127
  Office of Nuclear Security, 148
  Program 93+2, 125, 147, 148
  reference lab, 163
  safeguard agreements, 43, 45
  Safeguards System, 47, 76, 125–8
  Standing Advisory Group on Safeguards Implementation (SAGSI), 125
International Commission on Nuclear Non-proliferation and Disarmament (ICNND), 169, 171, 173
International Court of Justice (ICJ) case, 101, 132–6
International Nuclear Fuel Cycle Evaluation (INFCE), 80–82, 85, 194
International Security Outlook, 59
INVAP S.E., 163
Iran, 141, 149, 153, 156, 158, 160, 188
Iraq, 188
  invasion of, 152–3
  nuclear capability of, 109, 116, 125, 141, 149, 153
Ireland, 6
Israel, 55
Italy, 20, 55

Jabiluka uranium deposit, 49
Japan, 6, 25, 63, 71, 84, 107, 118, 171, 172, 173
  explosions in, 15
  Fukushima disaster, 177
  reactors in, 21, 125
  treaty ratification by, 55

uranium enrichment by, 42
uranium exports to, 43
Jervis Bay reactor, 35, 43–4
Jindalee Over-the-Horizon-Radar (JORN), 124
Joint Committee on Foreign Affairs and Defence (JCFD), 96, 98
Joint Facilities, 59–60, 93, 94, 95, 119, 124, 152
Joint Standing Committee on Treaties, 180

Kawaguchi, Yoriko, 171
Kazakhstan, 21
KazAtomProm, 21
Keating, Paul, 77, 115, 134–5, 136, 137
Keating government, 4, 9–10, 115–39, 161
disarmament efforts during, 131–8
geostrategic uncertainty during, 116–21
global nuclear diplomacy during, 128–31
non-proliferation efforts during, 121–5, 190
safeguards reform during, 125–8
Kelly, Paul, 141
Killen, James, 57–8, 59
Koongarra uranium deposit, 49
Kyoto Protocol, 169

latency, nuclear, 27
Leah, Christine, 8
Liberal-National Coalition, 141
Life-Cycle Extension Program (LEAP), 176
liquid metal fast breeder reactor (LMFBR), 74
London Suppliers Group, 72
Lowy Institute for International Policy, 12
Lucas Heights, 40, 43, 126
Ludlum, Scott, 181
Lyon, Rod, 8

Manhattan project, 15, 31
Marshall, Gavin, 184
massive retaliation, 16, 33
McKinnon Review, 161
McMahon Coalition government, 46
Medcalf, Rory, 175
medical isotopes, 19, 26, 163
Medvedev, Dmitry, 180
Meitner, Lise, 14
Member State Safeguard Support Program, 28
Menzies, Robert, 37, 39, 42
Menzies government, 32, 33, 34–5, 38
Millar, T.B., 57
minimum deterrence, 17
missile defence, 92, 94, 124–5, 145, 151, 156
mixed-oxide (MOX) fuels, 21
MOATA reactor, 40
molybdenum-99, 19, 26
Mott, Charles, 48

Nabarlek uranium deposit, 49
Nagasaki, Japan, 15
NATO. *See* North Atlantic Treaty Organization (NATO)
NATO/Warsaw Pact military balance in Europe, 58
Natural Resources Defense Council, 12
neutron activation, 19
neutron bomb, 16
neutron radiation, 19
neutron transmutation doping, 19
New Zealand, 6, 63, 91, 116
NNWS. *See* Non-Nuclear Weapons States (NNWS)
Non-Aligned Movement (NAM), 63, 133
Non-Nuclear Weapons States (NNWS), 45, 63, 156
non-proliferation, 1, 3, 6, 25, 26, 53–4, 61–3, 64, 72, 121–5, 190, 194, 197

Non-Proliferation of Nuclear Weapons Treaty (NPT), 1, 22, 23, 37, 45–9, 54, 62, 79, 88, 122, 132, 156, 158, 187, 188, 190, 196
  Preparatory Committee (1980), 66
  ratification of, 5, 35–6, 46, 48, 55, 91, 95, 115
  Review and Extension Conference (1995), 123, 125, 128, 130
  Review Conference (1975), 63
  Review Conference (1980), 65, 67, 68, 100, 107
  Review Conference (1985), 106, 107, 108
  Review Conference (1990), 107, 108
  Review Conference (2000), 150
  Review Conference (2005), 156
  Review Conference (2010), 171, 173
normative domain, 1, 22–4, 28
norm entrepreneurs, 22
North Atlantic Treaty Organization (NATO), 16, 18, 34, 68, 120
North Korea, 148, 153, 154, 156, 172, 188
  nuclear capability of, 116, 141, 149
  strategic concerns, 17
Norway, 6
NPT. *See* Non-Proliferation of Nuclear Weapons Treaty (NPT)
NSG. *See* Nuclear Suppliers Group (NSG)
Nuclear Disarmament Party, 100
nuclear energy policy, 12
Nuclear Expertise Steering Group, 161
nuclear fission policy, 13
Nuclear Industry Association, 12
Nuclear Non-Proliferation (Safeguards) Act, 112
nuclear policy, 1, 9, 11–29
  Australian, 3–7, 187–98
  characteristics of, 188–91
  definition, 11–14
  economic domain of, 18–22
  future of, 195–8
  normative domain of, 22–4
  strategic domain of, 14–18
  unifying elements of, 24–9
nuclear power policy, 12
Nuclear Security Summit, 170
Nuclear Suppliers Group (NSG), 23, 108, 109, 115, 182, 183, 188
nuclear weapons-free zones, 24
nuclear weapons policy, 11–12
Nuclear Weapons States (NWS), 1, 45, 156, 190
Nurrungar, 59, 95
NWS. *See* Nuclear Weapons States (NWS)
Nye Report, 118

Obama, Barack, 171, 172
Obama administration (US), 183
Oliphant, Mark, 31
Open-pool Light Water Reactor (OPAL), 163

Pacific Protector 03, 155
'Package of Practical Nuclear Disarmament and Non-Proliferation Measures', 173
Pakistan, 15, 55, 71, 72, 117, 160
  testing by, 149
  weapons in, 18, 141
peaceful nuclear explosion (PNE)
  in India, 17, 70, 72, 193–4
  in Soviet Union, 19
Peaceful Nuclear Explosions Treaty (1976), 100
Peacock, Andrew, 58, 64, 80, 85
Peko Mines Limited, 51
Peko-Wallsend, 84
PHWR. *See* Pressurized Heavy Water Reactors (PHWR)

Pine Gap, 59, 95
Plimsoll, Sir James, 65
plutonium-239, 13
plutonium economy, 73–80, 82, 193, 194
PNE. *See* peaceful nuclear explosion (PNE)
Portugal, 41
power, nuclear. *See* energy, nuclear
Pressurized Heavy Water Reactors (PHWR), 20
Pressurized Water Reactors (PWR), 20
Program 93+2, 125, 147, 148
PWR. *See* Pressurized Water Reactors (PWR)

radiation, neutron, 19
Ranger uranium deposit, 49, 50, 51
Ranger Uranium Environmental Inquiry, 51–3, 69, 75, 105
Rapid Deployment Force, 61
Rarotonga Treaty, 101–3
RBMK reactors, 20
reactors, nuclear, 19–22, 25–6
  in Australia, 35, 40, 43
  in Canada, 20, 41
  in China, 20, 21, 177
  in France, 20, 21, 25
  in Germany, 20, 125
  in India, 20, 21, 177
  in Japan, 21, 125
  liquid metal fast breeder reactor (LMFBR), 74
  manufacturers of, 21
  in Russia, 20, 21
  state involvement in, 25–6
  types of, 20
  in UK, 20
  in US, 20
Reagan, Ronald, 85, 86
Reagan administration (US), 91, 94, 100
Regional Security of Radioactive Sources (RSRS), 148
Reith, Peter, 152
Report of the Research Reactor Review (RRR), 161
research, 8, 39, 40, 163, 191, 197
Research School of Physical Sciences, 39
Reus-Smit, C., 68
Reynolds, Wayne, 80
Rice, Condoleezza, 157, 158
Richardson, J.L., 46
Rio Tinto, 21, 42
Robb, Andrew, 182
rogue states, 123, 151, 188
Rosatom, 21
Royal Australian Air Force (RAAF), 36
Rudd, Kevin, 167, 169, 170, 171, 175, 177, 178, 179, 181, 184
Rudd-Gillard-Rudd governments, 10, 169–86
  disarmament efforts during, 170–74
  extended deterrence during, 174–6
  relationship with India, 181–6
  uranium exports during, 177–81
Rudd government. *See* Rudd-Gillard-Rudd governments
Rumsfeld, Donald, 58, 151
Russia, 25, 141, 172
  reactors in, 20, 21
  testing by, 129
  uranium exports to, 180
  weapons in, 16

safety, nuclear, 24, 177
SALT. *See* Strategic Arms Limitation Talks (SALT)
SALT II, 23, 102
Saudi Arabia, 55
Scherger, Frederick, 36
Scholes, Gordon, 97, 98
Securing Australia's Energy Future (2004 White Paper), 164
security, nuclear, 7, 24, 26, 116–21, 190, 192, 197
Shoreham, Long Island, reactor, 25

Shultz, George P., 94
Sikkink, Kathryn, 7
SILEX Systems, 164–5
Sinclair, Ian, 75, 78, 79, 181
Smith, Stephen, 179–80, 182, 183, 184–5
So San, 154
South Africa, 18, 25, 41, 55, 115, 116
South Korea, 20, 25, 71
South Pacific Nuclear Free Zone (SPNFZ), 101, 103, 105, 109
South Pacific Nuclear Free Zone Treaty Act, 109, 112
Soviet Union, 62, 68, 71, 93, 115
  peaceful nuclear explosions in, 19
  testing by, 129
  weapons in, 16, 56, 57, 58–9
SPNFZ. *See* South Pacific Nuclear Free Zone (SPNFZ)
Spooner, W.H., 42
SSAC. *See* State System of Accounting for and Control of Nuclear Material (SSAC)
stability-instability paradox, 15, 59
START. *See* Strategic Arms Reduction Treaty (START)
START II, 23, 141
'Statement on Nuclear Policy' (Ford), 13–14
State System of Accounting for and Control of Nuclear Material (SSAC), 23
Stockpile Stewardship Program (SSP), 176
Strassmann, Fritz, 14
Strategic Arms Limitation Talks (SALT), 23, 56
Strategic Arms Reduction Treaty (START), 23, 115
Strategic Basis of Australian Defence Policy (1956), 33
Strategic Basis of Australian Defence Policy (1968), 95–6
Strategic Basis of Australian Defence Policy (1975), 57, 58
Strategic Basis of Australian Defence Policy (1983), 92, 97, 98
Strategic Concepts (1991), 120
Strategic Concepts (1999), 120
strategic domain, 1. *See also* weapons, nuclear
  nuclear policy, 14–18, 28
  state involvement in, 25, 26–7
strategic nuclear balance between the U.S. and the U.S.S.R., The, 58
Strategic Review 1993, 117, 119, 120, 122
Strategic Review 1997, 144–5
Strauss, Lewis, 40
Street, Tony, 68
Suitable Piece of Real Estate, A, 60
Suter, Keith, 63–4
Swan, Wayne, 184
Sweden, 6, 130
Switkowski, Dr. Ziggy, 166
Switzerland, 25, 55, 107
Sydney Declaration, 144
Syria, 141

Taiwan Straits Crisis, 118, 143–4
Tange, Sir Arthur, 34, 57
Tanter, Richard, 60
technetium-99, 19
Tenex, 21
terrorism, 151
testing, nuclear, 124, 129–30, 149
  by China, 130, 144, 145
  by France, 101, 129, 130
  by Soviet Union, 129
  by UK, 129
  by US, 60, 93–4, 124, 129, 152
thermonuclear warheads, 16
Threats to Australia's Security: Their nature and probability, 96–7
Three Mile Island, 21
Threshold Test Ban Treaty (1974), 100
Toshiba-Westinghouse, 21
Towle, P., 67–8
Treaty of Tlatelolco, 102

Trood, Russell, 81, 164
TVEL, 21

UEGA. *See* Uranium Enrichment Group of Australia (UEGA)
UMPNER. *See Uranium Mining, Processing, and Nuclear Energy Review* (UMPNER)
UN. *See* United Nations (UN)
Ungerer, Carl, 6
Union of Concerned Scientists, 12
United Arab Emirates (UAE), 181
United Kingdom (UK), 25, 71, 107–8, 130, 171
  nuclear capability of, 15, 41
  reactors in, 20
  relationship with Australia, 4, 34, 35
  testing by, 129
  weapons in, 16, 128
United Nations (UN)
  Atomic Energy Commission, 39, 41
  Conference on Disarmament, 64, 68, 100, 130, 131, 146
  Resolution 49/75K, 133
  Special Session on Disarmament, 64
United Nations General Assembly (UNGA), 45, 47, 64, 100, 133, 146
United States (US), 25, 62, 84, 107
  Atomic Energy Commission, 24, 40
  Atoms for Peace, 40
  'Commission to Assess the Ballistic Missile Threat to the US', 151
  counterproliferation initiative, 122
  Department of Defense, 11, 122–3
  Department of State, 11, 46–7
  Energy Research and Development Administration, 24
  Global Nuclear Energy Partnership (GNEP), 164
  Global Threat Reduction Initiative, 148
  National Missile Defence (NMD), 151
  nuclear capability of, 15, 41
  Nuclear Posture Review (NPR), 172, 173
  Nuclear Regulatory Commission, 24
  Proliferation Security Initiative (PSI), 23, 148, 154–5, 156, 159, 190
  reactors in, 20
  relationship with Australia, 4, 34–5, 56, 118–19, 143–4, 150, 152, 156, 191–2
  Strategic Defense Initiative (SDI), 92, 94, 124
  strategic policy of, 16–18
  testing by, 60, 93–4, 124, 129, 152
  treaty ratification by, 149, 150
  weapons in, 15, 16
United States Security Strategy for the East Asia–Pacific Region, 118
University of Western Australia, 127
uranium. *See also* exports of uranium
  Australian resources of, 21, 31, 49–50, 78, 193
  conversion of, 21
  enrichment of, 21, 50, 70, 83–7, 104, 164–6
  mining of, 21, 40, 51, 104, 178, 179
Uranium, the Joint Facilities, Disarmament and Peace, 95, 134
uranium-235, 13
Uranium Advisory Council, 104
Uranium Enrichment Group of Australia (UEGA), 84, 87, 104
Uranium Mining, Processing, and Nuclear Energy Review (UMPNER), 166–7, 169, 191
URENCO, 21, 84, 87, 166
USSR. *See* Soviet Union

Walker, William, 22
Waller, Keith, 46
Walsh, Jim, 31–2
Walsh, Peter, 104
Ward, Dr R.G., 85, 86
weapons, nuclear, 8, 11–12, 15–18, 25, 190
  Australian development of, 31–8, 44, 95–9, 192, 196–7
  in China, 16, 17, 35, 62, 128
  in France, 16, 62, 128
  in Pakistan, 18, 141
  in Russia, 16
  in the Soviet Union, 16, 56, 57, 58–9
  in the UK, 16, 128
  in the US, 15, 16
  use of, 15
weapons of mass destruction, 116, 123, 153–5
Weapons of Mass Destruction (Prevention of Proliferation) Act (1994), 124
Weinberger, Caspar, 94
Wesley, Michael, 7–8
West Germany, 63, 71, 72, 107
Westinghouse, 21
White, Hugh, 120
Whitlam, Gough, 48
Whitlam government, 5, 46, 48–51, 60
WMC, 84
World Health Organization (WHO), 132–3
World Nuclear News, 12
World War II, 15

yellowcake. *See* uranium
yttrium-90, 19

For Product Safety Concerns and Information please contact our EU representative GPSR@taylorandfrancis.com
Taylor & Francis Verlag GmbH, Kaufingerstraße 24, 80331 München, Germany

www.ingramcontent.com/pod-product-compliance
Ingram Content Group UK Ltd.
Pitfield, Milton Keynes, MK11 3LW, UK
UKHW011002180926
475701UK00004B/517

* 9 7 8 1 1 3 8 3 8 4 0 3 3 *